THE OPEN UNIVERSITY
A SCIENCE FOUNDATION COURSE

UNIT 15 CHEMICAL EQUILIBRIUM

THE OPEN UNIVERSITY PRESS

The Open University Press, Walton Hall, Milton Keynes, MK7 6AA.

First Published 1988. Reprinted 1990.

Designed by the Graphic Design Group of the Open University.

Filmset by Santype International Limited, Salisbury, Wiltshire; printed by Thomson Litho, East Kilbride, Scotland.

ISBN 0 335 16332 7

This text forms part of an Open University Course. For general availability of supporting material referred to in this text please write to: Open University Educational Enterprises Limited, 12 Cofferidge Close, Stony Stratford, Milton Keynes, MK11 1BY, Great Britain.

Further information on Open University Courses may be obtained from the Admissions Office, The Open University, P.O. Box 48, Walton Hall, Milton Keynes, MK7 6AB.

1.2

STUDY GUIDE

This Unit consists of three components: the text, two experiments and a TV programme. The two experiments are integrated with the text, one in Section 2, and one in Section 4.2. In both cases, the manipulations are very simple and each experiment should take about 40 minutes. The best results are obtained by using distilled or deionized water, which can be bought from any good chemist or garage. One litre should be adequate.

The TV programme, 'Equilibrium rules—OK?', supports Sections 4 and 5, and you will gain most benefit from it if you have studied up to the end of Section 4.5 before viewing the programme. Notes on the programme are in Section 8.

You will also be able to use a computer-assisted learning (CALCHEM) program via the terminal at your local Study Centre. This program provides exercises and examples related to Sections 4 and 5.

1 INTRODUCTION

Let us begin this Unit with a thought experiment, using the elements hydrogen, chlorine and iodine which you met in Units 13–14.

At room temperature, hydrogen and chlorine are gases that consist of diatomic molecules. Suppose we mix 1.2 moles of $H_2(g)$ and 1.0 mole of $Cl_2(g)$ in a glass bulb at room temperature. The colourless hydrogen and yellow-green chlorine react to form another colourless gas, hydrogen chloride:

$$H_2(g) + Cl_2(g) = 2HCl(g) \qquad (1)$$

Eventually the reaction stops, and there is no further change. Suppose we now find out what the bulb contains.

□ What substances would you expect to find, and in what amounts?

■ According to Equation 1, the 1.0 mole of Cl_2 will react with 1.0 mole of H_2 to give 2.0 moles of HCl(g). Apart from this HCl, there should just be 0.2 mole of unchanged H_2. *No chlorine should be present.*

This answer is correct, but it is made possible by an assumption. We have assumed that Reaction 1 is complete: that it proceeds until, *as far as we can detect*, one of the reactants has been completely used up. You made the same assumption about the tin–iodine reaction in Experiment 2 of Units 13–14. In many cases this assumption is very satisfactory, but we shall now look at an instance where it is not.

Suppose we repeat the hydrogen–chlorine experiment with 1.2 moles of H_2 and 1.0 mole of I_2. Iodine is a solid at room temperature, so to make both the reactants gaseous, as they were before, we will heat the bulb to, say, 427 °C. At this temperature, iodine is a purple vapour. The purple colour fades as the iodine reacts with the hydrogen to form the colourless gas, hydrogen iodide:

$$H_2(g) + I_2(g) = 2HI(g) \qquad (2)$$

This time however, the reaction stops before all the iodine colour has faded, and no matter how long we wait, it will not disappear. At this point, the bulb contains about 0.35 mole of H_2, 0.15 mole of I_2 and 1.70 moles of HI.

In this case, therefore, the reaction does not go to completion; measurement shows that, when the reaction is over, all the substances on both the left-hand and right-hand sides of Equation 2 are detectable. In such cases, we say that there is a state of balance, or *chemical equilibrium*, between the two sides of the equation, between the reactant(s) and the product(s).

If chemical equilibrium decides the extent of a chemical reaction, chemists and the chemical industry need to be thoroughly acquainted with it. In

ACID

BASE

INDICATOR

Units 15 and 16, we explore the nature of chemical equilibrium using illustrative examples. For two reasons, our examples will mainly involve the reactions of ions in aqueous solution: first, such reactions can often be performed with simple chemicals and glassware; secondly, they are often fairly quick, dramatic and important. From this Unit for instance, you can work out why orange juice is more palatable than vinegar, and why power stations increase the acidity of rainwater. However, before you can tackle either chemical equilibrium, or these questions, you must first consolidate and extend what you already know about aqueous ions, and this is the purpose of Sections 2 and 3.

In Experiment 4 of Units 13–14, you classified aqueous solutions by their electrical conductivity. Now we shall classify them by studying their effect on a substance called litmus. This is a dye extracted from the litmus lichen (*Roccella tinctoria*), which grows on cliffs along the coasts of the Canary Islands, and southern and western Europe. Surprisingly enough, we shall then find that the two kinds of classification can be related.

2 ACIDS AND BASES (Experiment 1)

Today, chemists are wary of the chemicals that they work with, but in the Middle Ages they were made of sterner stuff. For example, during that time, chemists noticed that the solutions of some substances possessed a sour taste whereas others tasted bitter and felt greasy when they were rubbed between finger and thumb. The first type of substance (or its solution) they called an **acid** (Latin, *acidus*, sour), and the second has come to be known as a **base**. A later and more complete classification of acids and bases is given below.

The aqueous solution of an acid:

- has a sour taste;
- is capable of changing the colour of many naturally occurring dyes. For example, it turns syrup of violets and many other blue vegetable substances red;
- in many cases, dissolves chalk to produce a gas. Chalk is calcium carbonate ($CaCO_3$) and the gas liberated when it is dissolved in an acid is carbon dioxide (CO_2);
- in many cases, dissolves metals such as magnesium or zinc to produce hydrogen gas (H_2).

The aqueous solution of a base:

- has a bitter taste;
- feels greasy;
- is capable of restoring the colour of many naturally occurring dyes after they have been changed by an acid.

In Experiment 1 we ask you to classify a number of chemical and household substances simply by testing their action on the dye litmus. DO NOT TASTE THEM! Dyes whose colours respond to the action of acids and bases are called **indicators**.

EXPERIMENT 1

CLASSIFICATION OF SOME CHEMICAL SUBSTANCES IN SOLUTION ACCORDING TO THEIR ACTION ON LITMUS

TIME
about 40 minutes

NON-KIT ITEMS
distilled water

KIT ITEMS

Chemical tray
ammonia solution ($2\,mol\,l^{-1}$)
dilute acetic acid ($2\,mol\,l^{-1}$)
dilute hydrochloric acid ($2.4\,mol\,l^{-1}$)
litmus papers, blue and red
litmus, solid
magnesium hydroxide
sodium carbonate
sodium chloride
sodium hydroxide

Tray A
dropping pipette
test-tubes (nine)
glass stirring rod

Tray C
measuring spoon (spatula-type)
spatula
test-tube brush
test-tube rack
wash bottle

Part 1

This part of the experiment is concerned with four solids, magnesium hydroxide, sodium hydroxide, sodium chloride and sodium carbonate. Add one of the solids to a clean, dry test-tube until the rounded bottom is just about full. (*Use too little rather than too much.*) Label the test-tube with the name of the solid. Now repeat this operation with the other three solids and three more test-tubes. Add distilled water from the wash bottle to a depth of 2 cm to the four test-tubes and shake to mix. Now tear up a red and a blue litmus paper into small pieces and add one red and one blue piece to each test-tube. If both papers become red, call the substance an acid; if both become blue, call it a base. If neither paper changes colour, classify the substance as 'neutral'. Put ticks in the appropriate columns of Table 1, which has already been completed for some other common substances. *Please remove the papers from the* $Mg(OH)_2$ *test-tube with the glass rod and retain the test-tube and contents for future work in Section 4.*

TABLE 1 Classification of some chemical and household substances in aqueous solution according to their action on litmus

Solute	Acid	Base	Neutral
SOLIDS			
calcium hydroxide		✓	
magnesium hydroxide			
sodium chloride			
sodium carbonate			
sodium hydroxide			
potassium nitrate			✓
sugar			✓
washing soda		✓	
LIQUIDS			
ammonia ($2\,mol\,l^{-1}$)			
acetic acid ($2\,mol\,l^{-1}$)			
hydrochloric acid ($2\,mol\,l^{-1}$)			
vinegar	✓		
lemon juice	✓		
orange juice	✓		
washing-up liquid		✓	
the 'air' you breathe out (containing carbon dioxide, CO_2)			

Part 2

This part of the experiment is concerned with the ammonia, acetic acid and hydrochloric acid solutions. Add two or three drops of your ammonia solution to about a 2 cm depth of distilled water in a fresh, clean test-tube. Test with litmus paper as in Part 1. Rinse the pipette with distilled water, and repeat this test with acetic acid instead of ammonia solution; then repeat with hydrochloric acid. Again put ticks in Table 1.

Part 3

Litmus also comes in a solid soluble form. Make up a solution of it by dissolving half a 'spoonful' in a 4 cm depth of distilled water. *Again, do not exceed this amount.* Divide the solution in two and blow through one part with a washed dropping pipette (a drinking straw will do equally as well). You may need to do this for up to five minutes to get

EXPERIMENT CONTINUED

a perceptible result. Does the part you blow through become more red or more blue when compared with the other solution that you prepared? Record your conclusion in Table 1.

Part 4

If you wish, you can check the entries for common household substances in Table 1 that we have completed for you by mixing the substances with distilled water.

Now clean your glassware with water, retaining just the $Mg(OH)_2$ test-tube as advised. You may wish to keep the glassware and chemicals to hand, since several of the items will be needed for Experiment 2.

NEUTRALIZATION

ARRHENIUS DEFINITIONS

The classification of Table 1 is consistent with household experience. Thus, like lemon juice and many other citrus fruit juices, vinegar tastes sour. As you probably know, vinegar is essentially a dilute aqueous solution of acetic acid, and the citrus fruits contain citric acid. Among the bases, you are probably familiar with the slippery feel of washing soda and soap solution, and with the unpleasantly bitter taste of the latter.

2.1 NEUTRALIZATION

Section 2 defined distinct, characteristic properties of acids and bases, but it did not explain why the two kinds of substance should be discussed under the same heading. One very good reason is their capacity for mutual destruction, a process which is called **neutralization**. Consider sodium hydroxide solution and hydrochloric acid. If they are mixed in the right amounts, the bitter taste of the base and the sour taste of the acid disappear, the mixture no longer changes the colour of litmus paper, and it tastes rather like seawater. The distinctive properties of an acid or a base disappear when one is added to the other. This type of reaction is called a neutralization reaction. Thus, in addition to its characteristic properties listed in Section 2, an acid reacts with a base to destroy or diminish its basic properties. Likewise, a base reacts with an acid to destroy or diminish its acidic properties.

2.2 INFLUENCE OF THE IONIC THEORY ON THE CONCEPTS 'ACID' AND 'BASE'

So far, the terms 'acid' and 'base' have been defined by the external effects that they have on other objects or materials (including each other, and ourselves!). This type of definition is said to be *operational*: it merely lists the operations that define the class of compounds. But these definitions provide no chemical reason why an acid or base possesses the characteristic properties that it does. Let us try to redeem this failing.

The hydrochloric acid that you used in Experiment 1 is made by dissolving gaseous hydrogen chloride (HCl) in water. If this solution is tested with the equipment used in Experiment 4 of Units 13–14 (Figure 1), the bulb lights up. Thus in water, HCl breaks down into ions, and in Section 7.6 of Units 13–14, we wrote the breakdown as:

$$HCl = H^+(aq) + Cl^-(aq) \quad (3)$$

Two other very well-known acids are nitric and sulphuric acids. They have the formulae HNO_3 and H_2SO_4, respectively, and their aqueous solutions also conduct electricity.

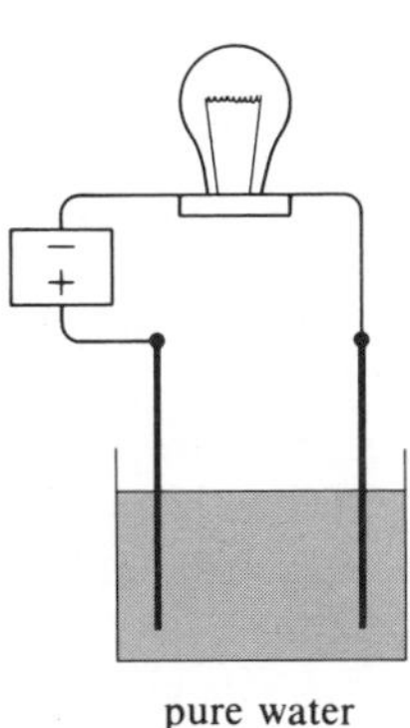

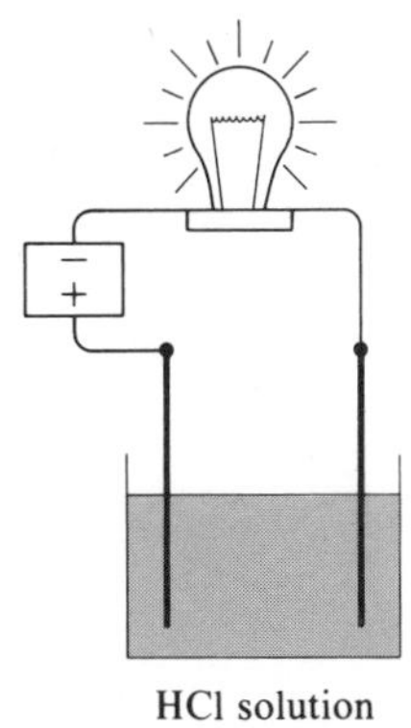

FIGURE 1 When 5 cm^3 of dilute hydrochloric acid are added to 50 cm^3 of water and tested as in Experiment 4 of Units 13–14, the bulb lights up. HCl is an electrolyte: its aqueous solution contains ions.

□ Write possible dissociation reactions for HNO_3 and H_2SO_4 in water.

■ Among the common aqueous ions listed in Table 5 of Units 13–14 are H^+, NO_3^- and SO_4^{2-}. Thus plausible dissociation reactions are:

$$HNO_3 = H^+(aq) + NO_3^-(aq) \quad (4)$$

$$H_2SO_4 = 2H^+(aq) + SO_4^{2-}(aq) \quad (5)$$

□ Look at Equations 3, 4 and 5. What do aqueous solutions of HCl, HNO_3 and H_2SO_4 have in common?

■ They contain a common ion, $H^+(aq)$.

Perhaps therefore, the common acidic properties of the three solutions are caused by the presence of $H^+(aq)$. If so, then *an acid can be defined as a substance that yields hydrogen ions,* $H^+(aq)$, *in aqueous solution.* Try ITQ 1 to convince yourself that this definition accounts for two of the characteristic properties of an acid.

ITQ 1 From Table 5 of Units 13–14, $Mg^{2+}(aq)$ and $Ca^{2+}(aq)$ are the ions that Mg and Ca are likely to form in aqueous solution. If an acid is simply a substance that yields $H^+(aq)$ ions in aqueous solution, write balanced equations that represent:

(a) the reaction between magnesium metal, Mg, and an aqueous acid, in which H_2 gas is evolved;

(b) the reaction between calcium carbonate, $CaCO_3$, and an aqueous acid, in which CO_2 gas and water are formed.

Now let us turn to bases. Consider two of these, sodium hydroxide, NaOH, and calcium hydroxide, $Ca(OH)_2$: when dissolved in water, both form conducting solutions.

□ Write equations for what happens when the two substances dissolve.

■ Among the aqueous ions listed in Table 5 of Units 13–14 are Na^+, Ca^{2+} and OH^-. Thus plausible dissociation reactions are:

$$NaOH(s) = Na^+(aq) + OH^-(aq) \quad (6)$$

$$Ca(OH)_2(s) = Ca^{2+}(aq) + 2OH^-(aq) \quad (7)$$

□ Can you now suggest a new definition of a base?

■ According to Equations 6 and 7, it seems likely that *a base is a substance that yields hydroxide ions,* $OH^-(aq)$, *in aqueous solution.*

These definitions of an acid and a base were first proposed in 1887 by the Swedish chemist, Svante Arrhenius, as part of his ionic theory for electrolyte solutions. They are known as the **Arrhenius definitions**, and they allow an elegant interpretation of the simple neutralization reactions discussed in Section 2.1.

SALT

2.3 NEUTRALIZATION AND THE ARRHENIUS DEFINITIONS

In Section 2.1, we discussed the reaction between *aqueous solutions* of solid sodium hydroxide and hydrogen chloride gas. When the reaction is carried out in the presence of much water, one cannot detect if water is produced or consumed in the reaction. Suppose, however, that we carry out this reaction between the pure substances without bothering to dissolve them in water. For instance, a slow stream of hydrogen chloride gas could be passed through a heated glass tube full of solid sodium hydroxide. What happens is that the sodium hydroxide is turned into solid sodium chloride (ordinary table salt) and *steam* comes out of the other end of the tube. The balanced equation for the reaction is:

$$NaOH(s) + HCl(g) = NaCl(s) + H_2O(g) \quad (8)$$

Thus by using the pure substances rather than the aqueous solutions, we see clearly that water molecules are an actual *product* of the reaction. This is not easily detected when the reaction happens in aqueous solution.

Now let us return to the reaction in aqueous solution at room temperature, and adjust Equation 8 accordingly. This time the reaction takes place between aqueous solutions of sodium hydroxide and hydrogen chloride. The products are sodium chloride and condensed steam (water). We could therefore write the reaction:

$$NaOH(aq) + HCl(aq) = NaCl(aq) + H_2O(l) \quad (9)$$

□ What criticism can be made of Equation 9?

■ In aqueous solution, NaOH, HCl, and NaCl consist of aqueous ions. This is not revealed by Equation 9.

□ Rewrite the equation so as to rectify this.

■ A reasonable answer is

$$Na^+(aq) + OH^-(aq) + H^+(aq) + Cl^-(aq)$$
$$= Na^+(aq) + Cl^-(aq) + H_2O(l) \quad (10)$$

However, this can be improved. The best kind of chemical equation gives us the *bare essentials of the chemical changes that occur*. In Equation 10, the ion $Na^+(aq)$ appears on both sides of the equation. This means that it is not changed by the reaction. Thus if it is eliminated from each side, the equation will still remain balanced, and still correctly describe the chemical change that has occurred.

□ Rewrite Equation 10 by eliminating the ions that occur on both sides.

■ By eliminating $Na^+(aq)$ and $Cl^-(aq)$, we get

$$H^+(aq) + OH^-(aq) = H_2O(l) \quad (11)$$

This result is very much simpler.

ITQ 2 Write the balanced equation that corresponds to Equation 10 for the neutralization reaction between potassium hydroxide, KOH, and nitric acid, HNO_3. Then improve the equation by eliminating the ions that occur on both sides.

ITQ 2 should have driven home the point that Equation 11 describes all simple neutralization reactions in water. The characteristic properties of an acid in aqueous solution are due to the presence of $H^+(aq)$ ions, whereas those of a base are caused by $OH^-(aq)$ ions. In neutralization reactions where acidic and basic properties suffer mutual annihilation, the $H^+(aq)$ and $OH^-(aq)$ ions combine to form water. Notice that this is one reason why acids and bases are important in many natural systems: they are linked through one of our vital compounds, water!

Equation 11 is a very elegant and general way of describing what happens in neutralization reactions. However, the alternative equation, Equation 9, does have one advantage: it tells us what solid substance we will get if, after the reaction has occurred, the products are dried out by evaporation of the water.

□ What is this solid substance?

■ Sodium chloride. The products in Equation 9 are water and an aqueous solution of sodium chloride: this explains the salty taste of the solution mentioned in Section 2.1. Loss of water leaves the solid.*

The overall result is that the base (sodium hydroxide) reacts with an acid (hydrochloric acid) to give water, which evaporates, and salt (sodium chloride), which contains the cation of the base and the anion of the acid. Because this reaction is the standard example of a large class of neutralization reactions, the word **salt** is now used as a general name for the solid residue left behind after the reaction between any acid and any base. Such neutralization reactions can be generally written as

$$\text{BASE} + \text{ACID} = \text{SALT} + \text{WATER} \qquad (12)$$

In such an equation the 'salt' is an ionic compound in which the cation, the positive ion, is derived from the base, and the anion, the negative ion, from the acid. Equations 11 and 12 are alternative ways of expressing the general reaction between a base and an acid in water. Equation 12 is helpful in identifying products, but it is Equation 11 that captures the essence of what happens.

ITQ 3 The solution produced by the reaction between KOH and HNO_3 in ITQ 2 is warmed so that the water evaporates. Name the solid residue, give its chemical formula, and state to what class of compounds it belongs.

2.4 THE ARRHENIUS DEFINITIONS: SOME LOOSE ENDS

Bases such as NaOH, $Mg(OH)_2$ and $Ca(OH)_2$ clearly contain the hydrogen and oxygen needed for the hydroxide ions that they form when they dissolve in water. Likewise, acids such as HCl, H_2SO_4 and HNO_3 contain the hydrogen needed for the formation of hydrogen ions. However, in Table 1, you should have found that solutions of ammonia (NH_3) and sodium carbonate (Na_2CO_3) were bases, and that a solution of CO_2 was slightly acidic. Yet NH_3 and Na_2CO_3 do not contain the required OH combination, and CO_2 lacks hydrogen. Can the Arrhenius definitions survive these discrepancies?

The answer is that they can, if we allow the solvent water molecules to participate in the equation that leads to the formation of $H^+(aq)$ or $OH^-(aq)$. For example, consider sodium carbonate, Na_2CO_3.

□ Write an equation for what happens when this solid dissolves in water.

■ $$Na_2CO_3(s) = 2Na^+(aq) + CO_3^{2-}(aq) \qquad (13)$$

So far, no $OH^-(aq)$ has appeared. However, in Table 5 of Units 13–14, we identified the bicarbonate ion, HCO_3^-, as a common anion in aqueous solution. When sodium carbonate dissolves in water, a small proportion of

* If you would like to try this experiment, first make up a solution of litmus as in Part 3 of Experiment 1. Dissolve two pellets of sodium hydroxide in about a 3 cm depth of water in a test-tube. Add two or three drops of your litmus solution, and then add dilute hydrochloric acid, drop by drop from a dropping pipette, shaking the solution as you do it, until the litmus *just* turns red. Pour the resulting solution into a saucer and leave it in a warm place. Inspect it when it has dried out.

CONCENTRATION

the $CO_3^{2-}(aq)$ on the right-hand side of Equation 13 reacts with water to form $HCO_3^-(aq)$:

$$CO_3^{2-}(aq) + H_2O(l) = HCO_3^-(aq) + OH^-(aq) \qquad (14)$$

In the process some $OH^-(aq)$ is formed, and this accounts for the basic properties of the solution.

☐ Write a similar equation that shows how dissolved CO_2, $CO_2(aq)$, can act as an acid.

■ Again a *little* of the CO_2 forms $HCO_3^-(aq)$:

$$CO_2(aq) + H_2O(l) = HCO_3^-(aq) + H^+(aq) \qquad (15)$$

☐ Now try to write an equation that shows how dissolved ammonia, $NH_3(aq)$, can act as a base

■ Table 5 of Units 13–14 lists the ammonium ion, NH_4^+, as a common cation in aqueous solution. We write

$$NH_3(aq) + H_2O(l) = NH_4^+(aq) + OH^-(aq) \qquad (16)$$

This reaction does indeed occur, although only to a limited extent: most dissolved ammonia remains as $NH_3(aq)$.

Thus by invoking Reactions 14, 15 and 16, which we emphasize occur only to a limited extent, the Arrhenius definitions can survive the three apparent exceptions.

SUMMARY OF SECTION 2

1 The aqueous solution of a typical acid tastes sour, turns blue litmus red, and dissolves magnesium, chalk and zinc. These characteristic properties are eliminated by the addition of a base.

2 The aqueous solution of a typical base tastes bitter, feels greasy and turns red litmus blue. These characteristic properties are eliminated by the addition of an acid.

3 These properties are explained by the Arrhenius definitions: acids yield $H^+(aq)$ in aqueous solution; bases yield $OH^-(aq)$ in aqueous solution.

4 The typical acid–base neutralization reaction can be written as

BASE + ACID = SALT + WATER

where the salt is a combination of the cation of the base with the anion of the acid. However, a more elegant description is

$$H^+(aq) + OH^-(aq) = H_2O(l)$$

SAQ 1 Tick to show *which and how many* of the words (a)–(d) refer to the statements (i)–(v).

	(a) an acid	(b) a base	(c) a salt	(d) water
(i) A substance that gives curdled milk its sour taste.	✓			
(ii) A substance that is formed in the reaction between aqueous solutions of an acid and a base.			✓	
(iii) A substance that turns red litmus blue		✓		
(iv) A substance that gives soap its slippery feel		✓		
(v) A substance that reacts with zinc to give hydrogen gas	✓			

SAQ 2 Write balanced equations of the type shown in Equation 10 for the neutralization reactions between: (i) aqueous solutions of calcium hydroxide, $Ca(OH)_2$, and nitric acid, HNO_3; (ii) aqueous solutions of lithium hydroxide, LiOH, and sulphuric acid, H_2SO_4.

Eliminate unchanged aqueous ions from both equations and show that they then reduce to Equation 11.

SAQ 3 Name and give the chemical formulae of the solids obtained by evaporating to dryness the solutions produced by the reactions in SAQ 2.

SAQ 4 Magnesium oxide, MgO, and sulphur trioxide, SO_3, are both solids. When added to water, MgO acts as a base, and SO_3 acts as an acid. Write balanced equations that show that this is consistent with the Arrhenius definitions. Table 5 of Units 13–14 may help you in selecting likely products.

3 THE CONCENTRATION OF A SOLUTION

Before you can study chemical equilibrium, you must understand one property of solutions that can be measured and expressed in numbers and units. Imagine two glasses of water. Suppose we dissolve one domestic sugar lump in one glass and two lumps in the other. The resulting solutions of sugar in water are different: the second is more *concentrated* than the first.

A useful way of describing such differences is to state the **concentration** of each solution. Concentrations can be specified in various ways. For instance, we could state the mass of dissolved sugar in a particular volume of solution.

ITQ 4 Suppose that each domestic sugar lump has a mass of 5 g, and that when the single sugar lump has dissolved in our first glass of water, the volume of the solution in the glass is $200\,cm^3$. What is the concentration of such a solution in grams of sugar per litre of solution ($g\,l^{-1}$)? (Remember 1 litre = $1\,000\,cm^3$.)

Specifying the mass of solute in a litre of solution is a perfectly valid way of expressing concentrations. However, as you saw in Units 13–14, chemical reactions are most conveniently thought of as reactions between small whole numbers of *moles* of substances. In chemistry, therefore, it is more useful to express concentrations as the number of *moles* of solute in one litre of solution (this is sometimes called the molarity of the solution).

ITQ 5 The chemical name of domestic sugar is sucrose, and it has the rather complex chemical formula $C_{12}H_{22}O_{11}$. What is the concentration of the solution of ITQ 4 in moles of sucrose per litre of solution ($mol\,l^{-1}$)? Use the relative atomic masses: C = 12; H = 1; O = 16.

The concentrations of solutions of ionic substances can also be expressed in this way, but in such cases something extra must be borne in mind. Suppose that 0.02 mole of magnesium chloride ($MgCl_2$) is dissolved in enough water to make one litre of solution. Obviously the concentration is $0.02\,mol\,l^{-1}$. However, from Section 6.3 of Units 13–14, you know that $MgCl_2$ dissociates in water:

$$MgCl_2(s) = Mg^{2+}(aq) + 2Cl^-(aq) \qquad (17)$$

We can state the concentration of the solution as 0.02 mole of $MgCl_2$ per litre because of the way the solution has been made. But we must remember that the dissolved $MgCl_2$ is present as $Mg^{2+}(aq)$ and $Cl^-(aq)$ ions.

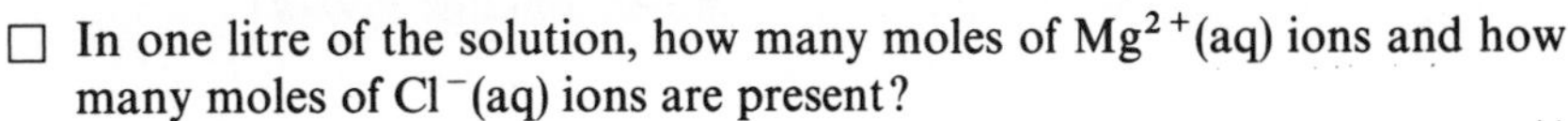
☐ In one litre of the solution, how many moles of $Mg^{2+}(aq)$ ions and how many moles of $Cl^-(aq)$ ions are present?

SATURATED SOLUTION

SOLUBILITY

CHEMICAL EQUILIBRIUM

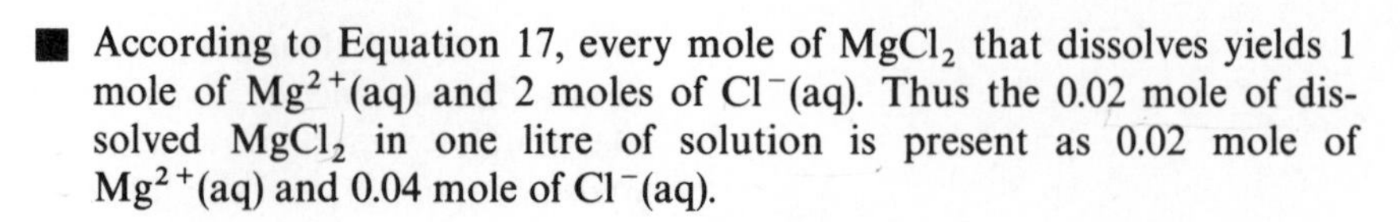

■ According to Equation 17, every mole of $MgCl_2$ that dissolves yields 1 mole of Mg^{2+}(aq) and 2 moles of Cl^-(aq). Thus the 0.02 mole of dissolved $MgCl_2$ in one litre of solution is present as 0.02 mole of Mg^{2+}(aq) and 0.04 mole of Cl^-(aq).

We could therefore describe the solution of $MgCl_2$ by saying that it contains 0.02 mol l^{-1} of Mg^{2+}(aq) and 0.04 mol l^{-1} of Cl^-(aq). However, it is customary to use the more concise description, which is derived from the way the solution was made, and to say that we are dealing with a solution of $MgCl_2$ of concentration 0.02 mol l^{-1}.

SAQ 5 500 cm^3 of a solution of sodium sulphate contains 7.1 g of dissolved Na_2SO_4. Calculate the concentration of (i) the solution, in mol l^{-1} of Na_2SO_4; (ii) Na^+(aq) in mol l^{-1}; (iii) $SO_4{}^{2-}$(aq) in mol l^{-1}. Use the approximate relative atomic masses: Na = 23; S = 32; O = 16.

SUMMARY OF SECTION 3

The amount of a chemical substance dissolved in a given volume of a solution is known as the concentration of that substance in the solution. Concentrations are usually expressed in moles per litre.

4 CHEMICAL EQUILIBRIUM

You are now ready to pursue the concept of chemical equilibrium, which was introduced in Section 1. To those who have not made an academic study of the term, the word 'equilibrium' implies a state of balance characterized by the appearance of quiescence: nothing seems to change as time passes. Such equilibrium is typified by a motionless object suspended from a spring (Figure 2a): this is a case of *mechanical* equilibrium. Another familiar example is the case of *hydrostatic* equilibrium shown in Figure 2b: a motionless block of wood floating in stagnant water. When we discuss chemical equilibrium, we shall occasionally pause, and compare it with such systems. As you will see, there are important similarities, and also important differences between chemical equilibrium and the other sorts. But first, we must consider a fresh example of chemical equilibrium through which the comparison can conveniently be made.

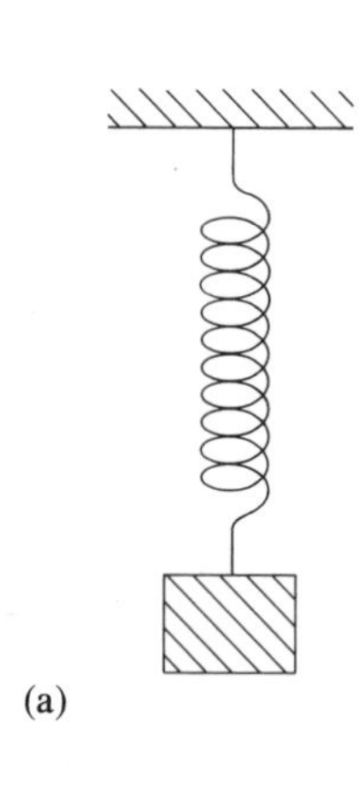

(a)

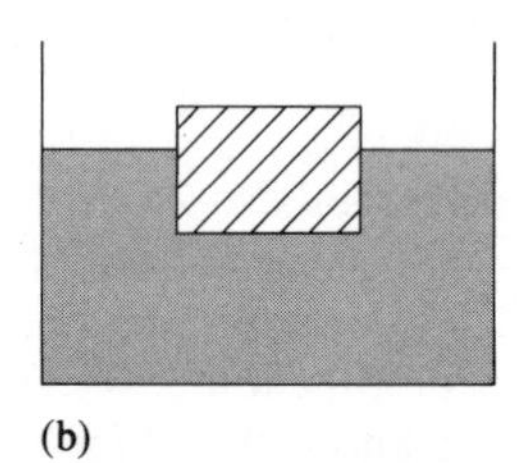

(b)

FIGURE 2 Two everyday instances of equilibrium: (a) a motionless object suspended from a spring—a case of mechanical equilibrium; (b) a block of wood floating in water—a case of hydrostatic equilibrium. Both examples have the appearance of quiescence: nothing seems to change with time.

4.1 A SATURATED SOLUTION: AN EXAMPLE OF CHEMICAL EQUILIBRIUM

A small amount of salt will dissolve easily in a cup of water. Add a little more, and that will dissolve as well. But this process cannot go on indefinitely: sooner or later, solid salt collects at the bottom of the cup, and no matter how hard you stir, or how long you wait (provided the temperature remains constant) no more will dissolve.

□ Can you suggest why we insist that the temperature stays constant?

■ The temperature is a factor that determines how much substance dissolves. Most solids, including sugar and salt, become more soluble in water as the temperature rises.

When, at a particular temperature, we reach the point when no more solute will dissolve in a solvent, the solution is said to be **saturated** at that temperature. The concentration of the solute in the saturated solution is then called the **solubility** of the solute in the solvent. For a particular combination of solute and solvent at a particular temperature, the solubility is a constant. For example, at 25 °C a saturated solution of sodium chloride in water contains 359.0 g l^{-1}.

Suppose we add a large amount of salt to some water at 25 °C and stir. To begin with, the solid dissolves to form aqueous sodium and chloride ions:

$$NaCl(s) \longrightarrow Na^{+}(aq) + Cl^{-}(aq) \quad (18)$$

In this equation, we have emphasized the direction of the dissolving process by replacing the usual equals sign with an arrow drawn from the reactant (solid NaCl) to the products (the aqueous ions). As more salt dissolves, the concentration of the solution rises and eventually reaches 359.0 $g\,l^{-1}$. At this point, the solution is saturated, and no more solid will dissolve. From then on, provided the temperature remains constant, the mass of solid salt at the bottom of the container, and the concentration of the solution above it, do not change with time. The system has 'the appearance of quiescence' referred to in Section 4, and the solid salt is said to be in **chemical equilibrium** with its saturated solution.

To summarize, the dissolving process, which in this case is represented by Equation 18, is limited: it can only go as far as the position of equilibrium allows. That position of equilibrium corresponds to the coexistence of undissolved solid with a saturated solution.

4.2 CHEMICAL EQUILIBRIUM—A STATE OF DYNAMIC BALANCE (Experiment 2)

You should have retained another example of a saturated solution from Experiment 1. This is a test-tube containing a suspension of solid magnesium hydroxide, $Mg(OH)_2$, in water.

When you made it up, you could not possibly have detected any dissolving of solid just by looking at it. This is because very little magnesium hydroxide dissolves—we say that magnesium hydroxide is very sparingly soluble in water. Nevertheless, there is some dissolution, and you subsequently found evidence for it.

□ What evidence was this?

■ In Experiment 1, you showed that the solution in contact with the solid turned red litmus paper blue. This suggests that some of the magnesium hydroxide has dissolved and produced aqueous hydroxide ions. The dissolution reaction is like that of $Ca(OH)_2$ in Equation 7:

$$Mg(OH)_2(s) \longrightarrow Mg^{2+}(aq) + 2OH^{-}(aq) \quad (19)$$

Obviously the solubility of magnesium hydroxide in water is much less than that of sodium chloride. At 25 °C, one litre of the saturated solution contains only 0.011 g of dissolved solid. Nevertheless, your test-tube is still an example of chemical equilibrium akin to the sodium chloride case discussed in Section 4.1. It contains solid magnesium hydroxide in equilibrium with aqueous magnesium and hydroxide ions.

Take a look at the test-tube. All should seem quiet. Given that the temperature is constant, neither the mass of the white solid, nor the concentration of the solution above it, is changing. The dissolving reaction, Equation 19, has apparently stopped. But this calm is deceptive. If you could climb into a frogman's outfit, shrink to the size of a water molecule and swim about in the test-tube, things would look very different. There is an experiment which can demonstrate this.

Many of the chemical elements around us are a blend of isotopes. This was demonstrated in Section 2.2.2 of Units 11–12. For example, the naturally occurring magnesium in your sample of $Mg(OH)_2$ consists of the isotopes ^{24}Mg, ^{25}Mg and ^{26}Mg. However, by using a nuclear reactor, it is possible to make magnesium hydroxide containing the isotope ^{28}Mg, and this isotope is radioactive. Chemically, this radioactive hydroxide behaves just like ordinary $Mg(OH)_2$, but the radioactive magnesium in it can be detected with a suitable device, such as a geiger counter.

Now suppose we add some of this solid $^{28}Mg(OH)_2$ to your test-tube. Once this solid has been added, there seems again to be no change: the amount of solid magnesium hydroxide does not diminish and the concentration of the saturated solution does not increase. This is what we would expect: $^{28}Mg(OH)_2$ behaves chemically just like natural $Mg(OH)_2$, and so the equilibrium is not disturbed. But if after a time, we filter off the solid, the residual solution is found to be radioactive due to the presence of $^{28}Mg^{2+}(aq)$. Now, following the addition of $^{28}Mg(OH)_2(s)$, there has been no increase in the total concentration of dissolved magnesium hydroxide.

□ So how can the presence of $^{28}Mg^{2+}(aq)$ in the solution be explained?

■ Some $^{28}Mg(OH)_2(s)$ must have dissolved, but this has been exactly balanced by the counter-movement of other isotopes of magnesium from the solution into the solid (Figure 3).

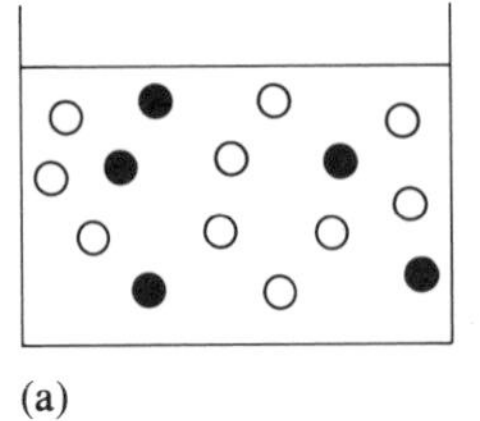

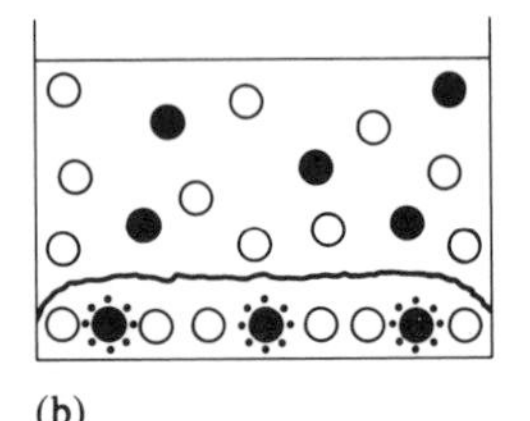

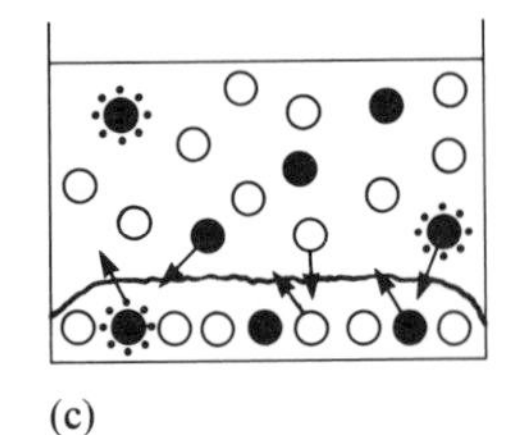

FIGURE 3 Radioactive labelling reveals the dynamic equilibrium that exists between a solid and its saturated solution. (a) A saturated solution of magnesium hydroxide: filled circles represent $Mg^{2+}(aq)$; open circles $OH^-(aq)$, the ratio being 1 : 2. (b) The same immediately after the addition of $^{28}Mg(OH)_2$: radioactive magnesium is represented by a halo of dots. (c) After some time: the concentration of the saturation solution is the same even though ^{28}Mg has moved into the solution because there is a counter-movement of non-radioactive magnesium into the solid.

At equilibrium, therefore, the reaction

$$Mg(OH)_2(s) \longrightarrow Mg^{2+}(aq) + 2OH^-(aq) \qquad (19)^*$$

proceeds just as it did before equilibrium was reached. The reverse reaction

$$Mg^{2+}(aq) + 2OH^-(aq) \longrightarrow Mg(OH)_2(s) \qquad (20)$$

however is also proceeding at an equal rate, and there is no net transfer of magnesium hydroxide in either direction. Beneath the apparent quiescence of the equilibrium state, there is, at the level of ions and molecules, a ceaseless coming and going. Chemical equilibrium is a *dynamic* process in which the forward (left → right) and reverse (right → left) reactions are going on at equal rates. To represent this, we shall write equilibrium systems with two half-headed arrows pointing in opposite directions. Thus a saturated solution of magnesium hydroxide is represented by the equation

$$Mg(OH)_2(s) \rightleftharpoons Mg^{2+}(aq) + 2OH^-(aq) \qquad (21)$$

Notice that although the hydrostatic equilibrium in Figure 2b resembles chemical equilibrium in its quiescence, it does not have this dynamic quality. This is a property where the analogy between the two systems breaks down.

You should now do Experiment 2. It consists of three parts, and provides practical demonstrations of the theoretical discussion still to come. Part 1 drives home the two-directional quality of Equilibrium 21 in a fresh way; the results are discussed in Section 4.3. Parts 2 and 3 illustrate the way in which the position of chemical equilibrium can be dramatically shifted by adding certain other chemicals. Part 2 works on Equilibrium 21, and the results are discussed in Section 4.4. Part 3 works on a new example of chemical equilibrium which will be of major concern throughout the rest of the Unit.

Record the results of the Experiment in your Notebook. You will need to refer to them later.

EXPERIMENT 2

CHEMICAL EQUILIBRIUM AND LE CHATELIER'S PRINCIPLE

TIME

about 40 minutes

NON-KIT ITEMS

distilled water

clean dry saucers (three)

KIT ITEMS

Chemical tray

dilute acetic acid (2 mol l^{-1})

dilute hydrochloric acid (2.4 mol l^{-1})

magnesium chloride

magnesium hydroxide

sodium acetate

sodium hydroxide

Tray A

glass stirring rod

test-tubes (eight)

Tray B

measuring cylinder, 25 cm^3

Tray C

measuring spoon (spatula-type)

test-tube brush

test-tube rack

wash bottle

Part 1

Fill the rounded bottom of one test-tube with magnesium chloride, and add two pellets of sodium hydroxide to another. Add distilled water from the wash bottle to both test-tubes to a depth of about 2 cm, and shake or stir until the solids have dissolved. Now pour the contents of one test-tube into the other. Note what happens in your Notebook.

Part 2

Stir the contents of the test-tube from Part 1 for about 30 seconds, and allow the solid to settle. Carefully tilt the tube to pour away as much clear liquid as possible, while leaving the solid in the test-tube. Now pour 5 cm^3 of dilute hydrochloric acid into the measuring cylinder. Add it, approximately 1 cm^3 at a time to the test-tube, shaking or stirring between the successive additions of acid. Note what happens in your Notebook.

(You may also try this experiment with the suspension of magnesium hydroxide in its saturated solution which you retained from Experiment 1. However, because the suspension is not freshly prepared, it may react rather slowly.)

Part 3

Fill the rounded bottom of a test-tube with sodium acetate, and add distilled water to a depth of about 2 cm. Shake until the solid dissolves. To two other test-tubes, add a 2 cm depth of dilute acetic acid. To one of these two tubes, add two pellets of sodium hydroxide, and shake until the pellets dissolve. Now pour the contents of each test-tube on to a clean, dry saucer. Compare the smell of the acetic acid that has been treated with sodium hydroxide with that of untreated acetic acid and with that of sodium acetate solution.

Your results should raise several questions. For instance, what is the substance produced in Part 1? This is answered in Section 4.3. The changes observed in Parts 2 and 3 can be thought of as a shift in the positions of certain chemical equilibria. What equations describe these equilibria, and why do the shifts occur in the particular directions that they do? These questions are dealt with in Sections 4.4 and 4.5.

4.3 TRYING TO BEAT THE EQUILIBRIUM SYSTEM

The amount of magnesium hydroxide that we can dissolve in water is limited by the equilibrium in Equation 21. And as we have seen, this equilibrium lies well over to the left-hand side of the equation: magnesium hydroxide is very sparingly soluble at 25 °C, and only very low concentrations of $Mg^{2+}(aq)$ and $OH^-(aq)$ coexist in the saturated solution. A solution containing higher concentrations of *both* these ions seems to be unobtainable, but we shall make one last attempt to get it.

Unlike $Mg(OH)_2$, both $MgCl_2$ and NaOH are very soluble in water, and you have already seen that their solutions contain ions:

$$MgCl_2(s) = Mg^{2+}(aq) + 2Cl^-(aq) \qquad (17)^*$$

$$NaOH(s) = Na^+(aq) + OH^-(aq) \qquad (6)^*$$

PRECIPITATE

LE CHATELIER'S PRINCIPLE

So by dissolving $MgCl_2$ in water, you can obtain a solution containing a high concentration of $Mg^{2+}(aq)$. Likewise, from NaOH, you can obtain a solution containing a high concentration of $OH^-(aq)$. If you now mix the two, the result will be a solution in which high concentrations of $Mg^{2+}(aq)$ and $OH^-(aq)$ exist together. These concentrations will be far higher than those obtainable by dissolving $Mg(OH)_2(s)$ in water, far higher, in other words, than Equilibrium 21 allows.

□ What happened when you tried this experiment?

■ A white solid suddenly appears when the two clear solutions are mixed (Experiment 2 Part 1).

When, as a result of a chemical reaction, a solid appears in what was previously a clear solution, the solid is called a **precipitate**. In this case the precipitate is magnesium hydroxide and it shows that you cannot beat the equilibrium system in this way. What happens is that, at these high concentrations, $Mg^{2+}(aq)$ and $OH^-(aq)$ combine to form $Mg(OH)_2(s)$:

$$Mg^{2+}(aq) + 2OH^-(aq) \longrightarrow Mg(OH)_2(s) \qquad (20)^*$$

The concentrations therefore drop until they are comparable with those obtained by dissolving $Mg(OH)_2(s)$ in water. At this point, precipitation of $Mg(OH)_2$ stops, and the chemical equilibrium described by Equation 21 once more exists. Throughout this drama, the other ions present in your mixed solutions, $Na^+(aq)$ and $Cl^-(aq)$, remain in solution because, unlike $Mg(OH)_2$, NaCl is freely soluble in water.

There are two general lessons to be learnt from this experiment. First, we can make a saturated solution of magnesium hydroxide either by allowing the $Mg(OH)_2(s)$ on the left of Equation 21 to dissolve in water, or by making the $Mg^{2+}(aq)$ and $OH^-(aq)$ on the right combine to form a precipitate. Thus chemical equilibrium, the state of balance between the two sides of the chemical equation, can be approached from two opposed directions: we can start from either side of the equation. As we saw in Section 4.2, at equilibrium the rates of the two opposed processes are equal and there is no net transfer of material from one side of the equation to the other.

Secondly, you now know that if, by mixing solutions of two readily soluble ionic compounds, you mix the ions derived from a very sparingly soluble compound, that compound will be precipitated.

Finally, notice that the equilibrium position in our hydrostatic analogy can, like chemical equilibrium, be approached from two opposed directions (Figure 4). The wooden block can either be let fall, or submerged and allowed to rise. However, as noted at the end of Section 4.2, the final equilibrium state does *not* have the dynamic quality of chemical equilibrium. There, the analogy breaks down.

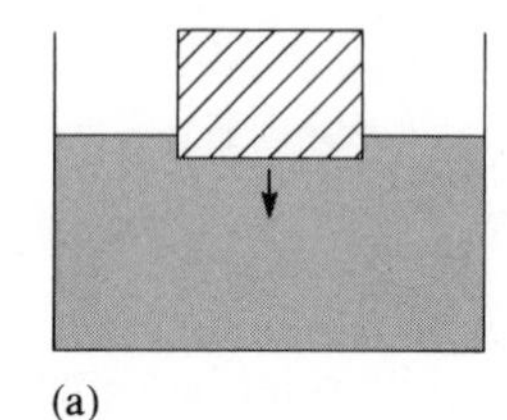

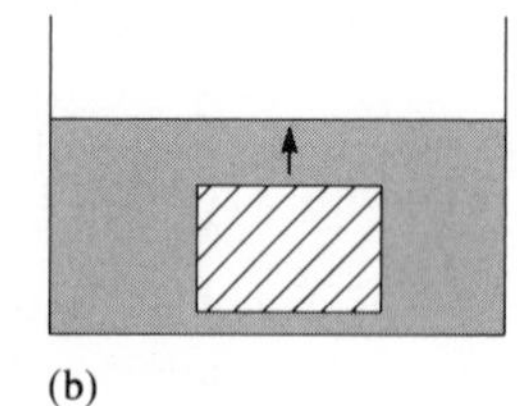

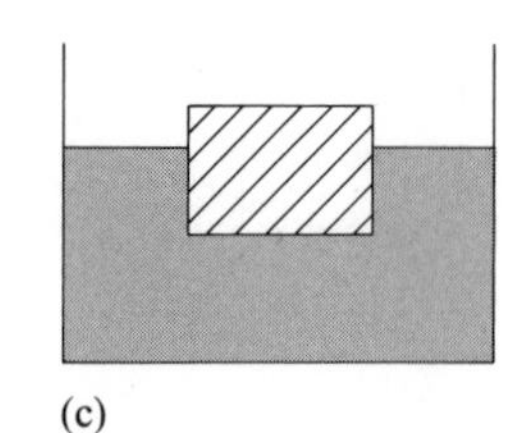

FIGURE 4 Like chemical equilibrium, this hydrostatic equilibrium (c) can be approached from two opposed directions. The wooden block can be let fall (a), or submerged and allowed to rise (b). But the equilibrium state (c) does *not* have the dynamic quality of chemical equilibrium.

4.4 LE CHATELIER'S PRINCIPLE

You now know that Part 1 of Experiment 2 yielded a precipitate of magnesium hydroxide in contact with its aqueous solution. Part 2 of Experiment 2 showed that when hydrochloric acid is added to this precipitate, the solid hydroxide dissolves. There is an especially illuminating way of looking at this change. Before the acid was added, we had the equilibrium

$$Mg(OH)_2(s) \rightleftharpoons Mg^{2+}(aq) + 2OH^-(aq) \qquad (21)^*$$

We argue that the addition of acid then *disturbs* this equilibrium, and that the subsequent dissolving of $Mg(OH)_2(s)$ is a *response to the disturbance.*

In what way does the addition of acid disturb Equilibrium 21? From Section 2.2 you know that acids are a source of $H^+(aq)$, which can react with one of the ions in Equation 21.

☐ Which one?

■ $H^+(aq)$ combines with $OH^-(aq)$ in the fundamental Arrhenius acid–base reaction:

$$H^+(aq) + OH^-(aq) = H_2O(l) \qquad (11)^*$$

This destruction of $OH^-(aq)$ lowers the concentration of this ion below the value that was present in the saturated solution of magnesium hydroxide: the equilibrium of Equation 21 is disturbed. How does the system respond to the disturbance? There is a principle that answers this question. It is called **Le Chatelier's principle**, after Henri Le Chatelier (1850–1936), a French mining engineer and chemist who first formulated it in 1888. We state it as follows:

> LE CHATELIER'S PRINCIPLE
>
> When a system in equilibrium is subjected to an external constraint, the system responds in a way that tends to lessen the effect of the constraint.

Let us use this principle to explain the effect of acid on Equilibrium 21.

☐ What external constraint does the equilibrium experience when $H^+(aq)$ is added?

■ As noted above, the concentration of $OH^-(aq)$, one of the ions present in Equilibrium 21, is lowered.

How can the system respond so as to lessen the effect of this constraint? At equilibrium, the right to left, and left to right reactions in Equation 21 proceed at equal rates. The lowering of the concentration of $OH^-(aq)$ throws the equilibrium out of balance and destroys this equality, so material must be transferred from one side of the equation to the other. Le Chatelier's principle tells us that this movement occurs from left to right: $Mg(OH)_2(s)$ dissolves and produces $OH^-(aq)$. It does so because, by this means, the external constraint—the loss of $OH^-(aq)$ caused by addition of acid—is lessened.

What net change is produced by all this activity? We started with Equilibrium 21 which has $2OH^-(aq)$ on the right-hand side. When hydrogen ions were added, we argued that they reacted with this '$2OH^-(aq)$' so we shall write the change by doubling Equation 11:

$$2H^+(aq) + 2OH^-(aq) = 2H_2O(l) \qquad (22)$$

As we have seen, Le Chatelier's principle tells us that this disturbance is followed by the net dissolution of more $Mg(OH)_2(s)$:

$$Mg(OH)_2(s) \longrightarrow Mg^{2+}(aq) + 2OH^-(aq) \qquad (20)^*$$

So the total change is the sum of Equations 22 and 20. Adding the two left-hand sides and two right-hand sides together separately, we get

$$Mg(OH)_2(s) + 2H^+(aq) + 2OH^-(aq) = Mg^{2+}(aq) + 2OH^-(aq) + 2H_2O(l) \qquad (23)$$

☐ Reduce this equation to a simpler form.

STRONG ELECTROLYTE

STRONG ACID

STRONG BASE

WEAK ELECTROLYTE

WEAK ACID

ACID STRENGTH

■ Following our treatment of Equation 10, we eliminate the $2OH^-(aq)$ on each side:

$$Mg(OH)_2(s) + 2H^+(aq) = Mg^{2+}(aq) + 2H_2O(l) \quad (24)$$

This equation is the best description of the overall change that occurs when acid is added to a suspension of $Mg(OH)_2(s)$ in water. As we have seen, it can be derived via an application of Le Chatelier's principle.

ITQ 6 Suppose that solid $MgCl_2$ is dissolved in a saturated solution of magnesium hydroxide, like the one you retained from Experiment 1. Measurements then show that the hydroxide ion concentration in the saturated solution is lowered. Explain this by applying Le Chatelier's principle.

Analogies of these applications of Le Chatelier's principle can be found in our hydrostatic equilibrium. Suppose that this time, the wooden block supports a small metal object (Figure 5a). At equilibrium the weight of metal and wood is exactly balanced by an upthrust or upward force from the water. Removal of the object (Figure 5b) is an external constraint. It disturbs the equilibrium by making the weight less than the upthrust. But the system now reacts to lessen the effect of the constraint: the block rises, and as it does so, the upthrust is lessened. Eventually weight and upthrust are once more in balance at a new equilibrium position (Figure 5c).

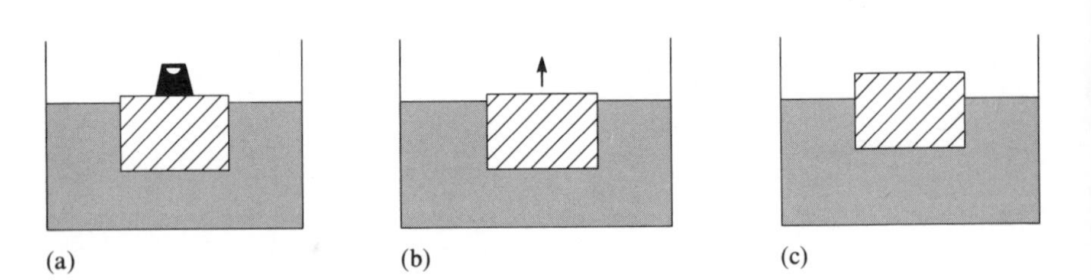

FIGURE 5 A system in hydrostatic equilibrium (a) is subjected to an external constraint by removal of the metal object (b). The block moves upwards to a new equilibrium position (c).

4.5 EQUILIBRIUM IN A SOLUTION OF A WEAK ACID

The two examples of chemical equilibrium that we discussed in Sections 4 to 4.4 were set up by adding electrolytes to water: both solid NaCl and $Mg(OH)_2$ dissolve and form aqueous ions. Sodium chloride and magnesium hydroxide, however, are electrolytes of a special kind, in that virtually *all* of the substance that dissolves breaks down into ions. Such substances are called **strong electrolytes**.

Most of the electrolytes that you have met so far have been of the strong variety. This is true, for example, of all salts, substances that, as we saw in Section 2.3, result from the combination of the cation of a base with the anion of an acid. Some acids, such as HCl and HNO_3, are also strong electrolytes: all the dissolved HCl and HNO_3 is broken up into ions, one variety of which is $H^+(aq)$:

$$HCl = H^+(aq) + Cl^-(aq) \quad (3)^*$$

Substances like HCl and HNO_3, which are both strong electrolytes and acids, are called **strong acids**. Likewise, substances like $Mg(OH)_2$ and NaOH are strong electrolytes by virtue of a dissociation which yields $OH^-(aq)$. They are called **strong bases**.

But not all acids and bases are strong electrolytes. To illustrate this, let us consider an experiment that you could do with the strong acid, HCl.

Earlier in this Unit (Figure 1), we noted that when some hydrochloric acid of concentration about $2.0\,mol\,l^{-1}$ is added to water, the conductivity of the water is greatly increased.

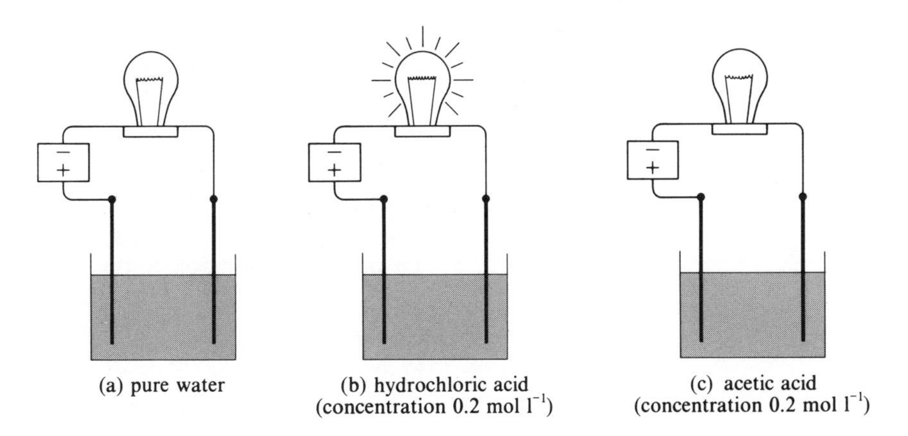

FIGURE 6 (a) Pure water is tested as in Experiment 4 of Units 13–14; (b) dilute hydrochloric acid is added until the acid concentration is about 0.2 mol l^{-1}: the bulb lights up; (c) the exercise is repeated with dilute acetic acid: the bulb remains unlit.

□ How was this increased conductivity detected?

■ The bulb in the electric circuit shown in Figure 6b lit up.

The greatly increased conductivity and the illumination of the bulb are made possible by the high concentration of ions in the HCl solution. This high concentration is due, in turn, to the fact that HCl is a strong acid: in solution it dissociates completely into ions. Now your Experiment Kit also contains acetic acid of concentration 2 mol l^{-1}. If you repeat the hydrochloric acid experiment with this acetic acid, you will find that this time the bulb does *not* light up (Figure 6c).

□ Does acetic acid solution contain any ions?

■ In Experiment 1, it turned blue litmus red: it must contain some H^+(aq).

□ Why then did the bulb not light up?

■ Unlike dissolved HCl, the dissolved acetic acid is *not* completely broken up into ions so the conductivity of water is not increased enough to illuminate the bulb.

Acetic acid has a complex formula $C_2H_4O_2$, which you will study further in Units 17–18. If it splits up in water into ions, one of the hydrogens breaks off as H^+(aq), and the rest of the formula is left as the acetate anion, $C_2H_3O_2^-$(aq). In this Unit, we shall keep things simple by using the symbol Ac for the collection of atoms $C_2H_3O_2$ (note that Ac is *not* a chemical element). Acetic acid is then denoted by HAc, thus emphasizing the single hydrogen that can be ionized in water, and the acetate ion becomes Ac^-(aq).

Now although acetic acid dissolves readily in water, the experiments in Figure 6 suggest that only a little of the dissolved acid breaks up into the ions H^+(aq) and Ac^-(aq). What we have in the solution is an equilibrium between the aqueous undissociated acid, and its aqueous ions:

$$\text{HAc(aq)} \rightleftharpoons \text{H}^+\text{(aq)} + \text{Ac}^-\text{(aq)} \quad (25)$$

Furthermore, the equilibrium inclines towards the left-hand side of this equation: most dissolved acid is in the form of neutral undissociated molecules, HAc(aq). Substances that are only *partly* broken down into ions when they dissolve in water are called **weak electrolytes**. If, like acetic acid, they are also acids, we call them **weak acids** to distinguish them from the strong acids, which are completely dissociated. (Weak bases also exist, but they are not discussed explicitly in this Course.)

Finally, in this context, note the important distinction between **strength** and **concentration**. A solution is concentrated if the concentration of the solute in the solution is high. An electrolyte is strong if, when dissolved, it is completely dissociated into ions. We can therefore have concentrated solutions of weak electrolytes: an acetic acid solution of concentration 10 mol l^{-1} is an example. Conversely, we can have dilute solutions of strong electrolytes: a saturated solution of magnesium hydroxide is dilute because the solid is only sparingly soluble in water, but what dissolves is completely dissociated into ions. So when talking about the concentration of a solution, it's a good idea to refer to 'dilute' or 'concentrated' solutions, rather than 'weak' or 'strong' solutions.

In Section 4.4, we explained the results of Part 2 of Experiment 2 by applying Le Chatelier's principle to the equilibrium of Equation 21. In ITQ 7, you are asked to explain the results of Part 3 of Experiment 2 by applying Le Chatelier's principle to the equilibrium of Equation 25.

ITQ 7 The characteristic smell of acetic acid is due to the presence of undissociated HAc molecules in the solution. Use Le Chatelier's principle to explain the following observations:

(i) When sodium hydroxide is dissolved in acetic acid, the acid's vinegar-like smell is much diminished;

(ii) When solid sodium acetate, NaAc, is dissolved in acetic acid solution, measurements show that the hydrogen ion concentration of the solution is greatly decreased.

SUMMARY OF SECTION 4

1 A system in chemical equilibrium appears to be quiescent at constant temperature.

2 Chemical equilibrium is dynamic: the appearance of quiescence is the result of two opposed chemical processes occurring at equal rates.

3 A state of chemical equilibrium can be approached from two opposed directions, corresponding to the two opposed processes.

4 When a system in chemical equilibrium is subjected to an external constraint, the system responds in a way that tends to lessen the effect of the constraint (Le Chatelier's principle).

5 All these points can be demonstrated by two important kinds of equilibrium system: the saturated solution of a strong electrolyte, and the equilibrium between the solution of a weak acid and its aqueous ions.

SAQ 6 Table 2 divides some lead (Pb), barium (Ba) and alkali metal (Na and K) salts into two categories: those that are readily soluble in water and those that are very sparingly soluble in water. The colours of the very sparingly soluble compounds are given in brackets.

TABLE 2 List of readily soluble and very sparingly soluble salts for SAQ 6. Colours of the very sparingly soluble salts are given in brackets. All of the freely soluble salts consist of colourless crystals

Freely soluble	Very sparingly soluble
$BaCl_2$, $Ba(NO_3)_2$,	$BaSO_4$(white)
Na_2CO_3, NaCl,	$BaCO_3$(white)
$NaNO_3$, Na_2SO_4,	$PbCl_2$(white)
KI, KNO_3,	PbI_2(yellow)
$Pb(NO_3)_2$	$PbCO_3$(white)
	$PbSO_4$(white)

State whether any visible chemical reaction occurs when aqueous solutions of the following compounds are mixed. If a reaction occurs, write an equation for it:

(i) Barium chloride, $BaCl_2$, and sodium sulphate, Na_2SO_4

(ii) Barium nitrate, $Ba(NO_3)_2$, and sodium chloride, NaCl

(iii) Lead nitrate, $Pb(NO_3)_2$, and potassium iodide, KI

(iv) Lead nitrate, $Pb(NO_3)_2$, and sodium carbonate, Na_2CO_3

SAQ 7 Hydrogen chloride gas, HCl, is dissolved in an aqueous solution of acetic acid. Will the concentrations of the acetate ion, $Ac^-(aq)$, and the undissociated acid, HAc(aq), rise or fall? (Hint: HCl is a strong electrolyte.)

SAQ 8 Write a balanced equation for the reaction between hydrogen gas, H_2, and nitrogen gas, N_2, to form gaseous ammonia, NH_3.

NH_3 is very soluble in water, but H_2 and N_2 are only slightly so. Suppose you have an equilibrium mixture of H_2, N_2 and NH_3, and the mixture is then washed with a spray of water. Use Le Chatelier's principle to predict what will happen.

SAQ 9 Assign the aqueous solutions (i)–(v) to the following categories: (a) a dilute solution of a weak electrolyte; (b) a concentrated solution of a weak or non-electrolyte; (c) a dilute solution of a strong electrolyte; (d) a concentrated solution of a strong electrolyte.

(i) Hydrochloric acid of concentration $12\,mol\,l^{-1}$

(ii) Acetic acid of concentration $0.002\,mol\,l^{-1}$

(iii) A saturated solution of magnesium hydroxide

(iv) A saturated solution of sugar

(v) A saturated solution of potassium iodide

5 THE EQUILIBRIUM CONSTANT

Let us look again at the experimental observations in ITQ 7. We began with a solution of acetic acid in which there exists the equilibrium

$$HAc(aq) \rightleftharpoons H^+(aq) + Ac^-(aq) \qquad (25)^*$$

We then increased the concentration of $Ac^-(aq)$ by dissolving solid sodium acetate in the solution. This threw the system out of balance, and material was transferred from one side of Equation 25 to the other.

□ How did this transfer affect the concentrations of HAc(aq) and $H^+(aq)$?

■ The system responded by trying to mitigate the increase in the concentration of $Ac^-(aq)$. Some $H^+(aq)$ therefore destroyed some $Ac^-(aq)$ by combining with it to form HAc(aq). Thus the concentration of $H^+(aq)$ was lowered, and that of HAc(aq) was increased. Then a new position of equilibrium was reached, and the changes stopped.

Notice that this explanation is not quantitative: it simply talks of increases and decreases. Can it therefore be expressed in a more quantitative way? The answer is that it can, because whenever HAc(aq) is in equilibrium with $H^+(aq)$ and $Ac^-(aq)$, there is a precise, numerical relationship between the concentrations of the three species. We shall begin by simply telling you what this relationship is.

Suppose we write the concentration of a species in the solution by enclosing the symbol for the species in square brackets. Thus the concentration of HAc(aq) is written [HAc(aq)]. Suppose also that we have a solution in which HAc(aq) is in equilibrium with $H^+(aq)$ and $Ac^-(aq)$:

$$HAc(aq) \rightleftharpoons H^+(aq) + Ac^-(aq) \qquad (25)^*$$

It turns out that there is a constant, K, given by

$$K = \frac{[H^+(aq)][Ac^-(aq)]}{[HAc(aq)]} \qquad (26)$$

At 25 °C, $K = 1.8 \times 10^{-5}\,mol\,l^{-1}$. This constant is called the **equilibrium constant** of the system in Equation 25. *In this case*, it is created by multiplying the concentrations of the two species on the right of the equation together, and dividing them by the concentration of the species on the left.

ACID DISSOCIATION CONSTANT

How can this help us to understand the response of the equilibrium to the increase in $[Ac^-(aq)]$? Because $[Ac^-(aq)]$ appears in the numerator of the expression on the right of Equation 26, the immediate result of the increase is to raise the value of this expression above the equilibrium value of $1.8 \times 10^{-5}\,mol\,l^{-1}$. This means that the system is no longer at equilibrium: to return to equilibrium, the value of the expression must be lowered back to $1.8 \times 10^{-5}\,mol\,l^{-1}$.

□ How can the transfer of material from the right to the left of Equation 25 achieve this?

■ If some $H^+(aq)$ and $Ac^-(aq)$ combine to form HAc(aq), the two concentrations in the numerator will decrease, and the single concentration in the denominator will increase. *All three changes lower the numerical value of the expression.* This process continues until the value is once more $1.8 \times 10^{-5}\,mol\,l^{-1}$.

In ITQ 7, we used Le Chatelier's principle to explain changes that occur when the concentration of a species in Equilibrium 25 is altered. You should now be able to see that these changes are a consequence of the requirement that Equation 26 must hold at equilibrium.

Finally, notice that in the particular case of Equation 26, the equilibrium constant has units of $mol\,l^{-1}$.

□ Can you see from Equation 26 why this is so?

■ Each individual concentration in Equation 26 is expressed in $mol\,l^{-1}$. As two concentrations are multiplied together on top, and then divided by a single concentration at the bottom, the units of K will be $mol\,l^{-1}$.

ITQ 8 You observed the effect of dissolved sodium hydroxide on Equilibrium 25 in Part 3 of Experiment 2: the smell of undissociated acetic acid molecules was diminished. Use Equation 26 to explain this result.

5.1 THE SIZE OF THE EQUILIBRIUM CONSTANT

You now know that when pure acetic acid is added to water, it forms undissociated molecules, HAc(aq), some of which then break up into ions:

$$\mathrm{HAc(aq)} = \mathrm{H^+(aq)} + \mathrm{Ac^-(aq)} \qquad (27)$$

This overall change continues until the equilibrium

$$\mathrm{HAc(aq)} \rightleftharpoons \mathrm{H^+(aq)} + \mathrm{Ac^-(aq)} \qquad (25)^*$$

is reached. Then at 25 °C,

$$\frac{[\mathrm{H^+(aq)}][\mathrm{Ac^-(aq)}]}{[\mathrm{HAc(aq)}]} = 1.8 \times 10^{-5}\,\mathrm{mol\,l^{-1}} \qquad (28)$$

Suppose that the amount of acetic acid that we add is just enough to make the final equilibrium concentration of *undissociated* acid, [HAc(aq)], equal to $1\,mol\,l^{-1}$. Let us work out what the equilibrium concentrations of $H^+(aq)$ and $Ac^-(aq)$ would be in this case.

□ First, what can you say about the *relative* values of $H^+(aq)$ and $Ac^-(aq)$?

■ Every HAc molecule that dissociates puts one $H^+(aq)$ and one $Ac^-(aq)$ into the solution. The numbers of the two kinds of ion, and therefore their concentrations, must be equal. Thus $[H^+(aq)] = [Ac^-(aq)]$.

We shall use the symbol y for this common concentration of $H^+(aq)$ and $Ac^-(aq)$. As $[HAc(aq)] = 1\,mol\,l^{-1}$, Equation 28 tells us that,

$$\frac{y \times y}{1\,\text{mol}\,\text{l}^{-1}} = 1.8 \times 10^{-5}\,\text{mol}\,\text{l}^{-1}$$

$$y^2 = 1.8 \times 10^{-5}\,\text{mol}^2\,\text{l}^{-2}$$

$$y = 4.2 \times 10^{-3}\,\text{mol}\,\text{l}^{-1}$$

Thus, every litre of the equilibrium solution contains 1 mole of HAc(aq), 0.0042 mole of H^+(aq) and 0.0042 mole of Ac^-(aq).

□ When the solution was first made up by adding acetic acid to water, how many moles of acetic acid were needed to make each litre of solution?

■ 1.0042 moles. 1 mole is needed to form the undissociated HAc(aq). Every molecule of HAc that dissociates forms one H^+(aq) and one Ac^-(aq). Therefore the residual 0.0042 mole is sufficient to form the 0.0042 mole of H^+(aq), and the 0.0042 mole of Ac^-(aq) in each litre.

The experiment pictured in Figure 6 suggested that very little of the dissolved acetic acid is dissociated into ions. We now have a quantitative reinforcement of this point. In the solution that we have just discussed, only about 0.4% of the dissolved acetic acid breaks up into ions. Reaction 27 does not proceed very far before the equilibrium of Equation 25 is reached: the equilibrium position lies well over to the left of Equation 25.

In the context of the calculation, we obtained a small percentage dissociation because K, the expression on the left of Equation 28, is very small. Study the calculation again to make sure you understand this; SAQ 10 at the end of this section provides further confirmation. Thus the size of K is an indication of how far Reaction 27 goes, that is, of the *strength* of this particular weak acid.

Now there are many other acids that, like acetic acid, break up in water into H^+(aq) and a singly charged anion, which we can write X^-(aq). As with acetic acid, in their aqueous solutions there exists the equilibrium

$$\text{HX(aq)} \rightleftharpoons \text{H}^+\text{(aq)} + \text{X}^-\text{(aq)} \tag{29}$$

for which we can write an equilibrium constant K_a,

$$K_a = \frac{[\text{H}^+\text{(aq)}][\text{X}^-\text{(aq)}]}{[\text{HX(aq)}]} \tag{30}$$

where K_a is an indication of how far the following reaction goes:

$$\text{HX(aq)} = \text{H}^+\text{(aq)} + \text{X}^-\text{(aq)} \tag{31}$$

This time we have added a subscript 'a' to K in recognition of the fact that we are dealing with the equilibrium constant of a particular kind of reaction—the dissociation of an acid. Not surprisingly then, K_a is called the **acid dissociation constant**.

Some values of K_a for acids of the type HX are shown in Table 3. Note first that hydrochloric and nitric acids are simply labelled 'strong acid'.

TABLE 3 Values of K_a for selected acids at 25 °C

Acid	Dissociation equilibrium	$K_a/\text{mol}\,\text{l}^{-1}$
hydrochloric	$HCl(aq) = H^+(aq) + Cl^-(aq)$	strong acid
nitric	$HNO_3(aq) = H^+(aq) + NO_3^-(aq)$	strong acid
iodic	$HIO_3(aq) \rightleftharpoons H^+(aq) + IO_3^-(aq)$	1.7×10^{-1}
nitrous	$HNO_2(aq) \rightleftharpoons H^+(aq) + NO_2^-(aq)$	4.5×10^{-4}
acetic	$HAc(aq) \rightleftharpoons H^+(aq) + Ac^-(aq)$	1.8×10^{-5}
hypochlorous	$HClO(aq) \rightleftharpoons H^+(aq) + ClO^-(aq)$	2.9×10^{-8}

□ Look back at Equation 30. Can you suggest why values of K_a are not quoted in these cases?

■ These acids are completely dissociated in aqueous solution: there is no evidence for any HX(aq). If [HX(aq)] is zero, K_a is infinite, so acid dissociation constants are not used with these strong electrolytes.

It is therefore meaningless to assign relative strengths to strong acids in aqueous solution: hydrochloric acid and nitric acid are equally strong. But the other K_a values in Table 3 are finite: these acids are weak acids, which are not completely dissociated into ions in aqueous solution, and their relative strengths can be compared in terms of the values of K_a. Thus iodic acid is stronger than nitrous acid, which is stronger than acetic acid.

SUMMARY OF SECTION 5

1 In an aqueous solution of an acid, HX, *at equilibrium*, the quantity

$$\frac{[H^+(aq)][X^-(aq)]}{[HX(aq)]}$$

always has a constant value, K_a, which is known as the acid dissociation constant or, most generally, as the equilibrium constant.

2 The way in which the weak acid system responds to changes in the concentration of HX(aq), H^+(aq), and X^-(aq) can be regarded as an attempt to maintain the quantity in question at the equilibrium value, K_a.

3 K_a is a measure of how far the following reaction proceeds before equilibrium is attained:

$$HX(aq) = H^+(aq) + X^-(aq)$$

SAQ 10 Iodic acid, HIO_3, is dissolved in water until the concentration of undissociated acid, $[HIO_3(aq)]$, is $1\,mol\,l^{-1}$. Use the K_a value in Table 3 to calculate (i) values for $[H^+(aq)]$ and $[IO_3^-(aq)]$, and (ii) the percentage of the total dissolved acid that is broken down into ions at 25 °C. (Hint: use your answer to (i) to work out the sum of the amounts of dissociated and undissociated acid in one litre of the solution.)

SAQ 11 0.1 mole of an acid HX is dissolved in water and the solution is made up to one litre. In this solution, half of the dissolved acid is in an undissociated state. What are the concentrations of HX(aq), H^+(aq) and X^-(aq) in the solution? What is the value of K_a for the acid?

Compare your answer with the value of K_a for nitrous acid (HNO_2) in Table 3. What do you conclude about the proportion of undissociated acid in an aqueous solution of HNO_2 of concentration $0.1\,mol\,l^{-1}$?

SAQ 12 Consider the solution of HNO_2 of concentration $0.1\,mol\,l^{-1}$ referred to in SAQ 11. Suppose it is compared with a solution of nitric acid, HNO_3, of concentration $0.05\,mol\,l^{-1}$; which of the following statements is correct?

(a) The HNO_2(aq) solution is the more concentrated.

(b) HNO_2(aq) is the stronger acid.

(c) $[H^+(aq)]$ is higher in the HNO_2(aq) solution.

6 WHAT ABOUT THE SOLVENT?

When, in Section 5.1, we used K_a to calculate the hydrogen ion concentration, we assumed that all the H^+(aq) originated from dissolved acid: water, the solvent, was regarded as a background medium which, by itself, supplies no hydrogen ions. This assumption is not correct, but in Section 5.1 that did not matter, because the amount of hydrogen ions provided by the water was negligible in comparison with the amount supplied by the acid. In solutions of much lower acidity this is not the case, and variations in hydrogen ion concentration cannot be understood without considering the contribution of the solvent. Solutions of this kind fall as rain, occur in rivers and seas, and circulate through plants and animals. Very small changes in their hydrogen ion concentration can have a profound and sometimes lethal effect on living things. Clearly then, we must try to understand how the solvent can influence hydrogen ion concentrations.

The assumption that, in problems of acidity, water can be regarded as mere background implies that pure water contains no $H^+(aq)$. This is consistent with Experiment 4 of Units 13–14: when water was tested with the circuit shown in Figure 6, the bulb did not light up. However, as Figure 6 shows, acetic acid solution does not light the bulb either, and we now know that it contains significant concentrations of ions. Perhaps therefore, any ion concentrations in pure water are too small to be detected by this relatively insensitive test. This idea is supported by common knowledge: water in contact with exposed electric wiring is dangerous because it can transmit electric shocks. Such conductivity supports the idea that water contains *some* ions.

□ What ions might be present in pure water?

■ $H^+(aq)$ and $OH^-(aq)$ are now very familiar to you. An obvious equation for the breakdown of a water molecule into ions is

$$H_2O(l) = H^+(aq) + OH^-(aq) \qquad (32)$$

The picture of pure water that all this conjures up is of a medium that is mainly water molecules, $H_2O(l)$, with small amounts of $H^+(aq)$ and $OH^-(aq)$ also present. In other words, in water there is an equilibrium:

$$H_2O(l) \rightleftharpoons H^+(aq) + OH^-(aq) \qquad (33)$$

The detectable, but very small conductivity of water suggests that only tiny amounts of $H^+(aq)$ and $OH^-(aq)$ are present. The left to right reaction of Equation 32 does not go very far so the equilibrium position in Equation 33 lies well over to the left. Since, from Section 4.3, we know that equilibrium can be approached from two opposed directions, this must mean that the reverse, right to left reaction must go a long way.

□ What evidence do you have that this is so?

■ We have, until now, regarded the fundamental Arrhenius acid–base reaction

$$H^+(aq) + OH^-(aq) = H_2O(l) \qquad (11)^*$$

as essentially complete, and this assumption has held good throughout the discussion so far.

6.1 THE ION PRODUCT OF WATER

Various pieces of evidence therefore support a picture of water as an equilibrium system,

$$H_2O(l) \rightleftharpoons H^+(aq) + OH^-(aq) \qquad (33)^*$$

in which equilibrium lies far to the left. This equilibrium corresponds closely to that for a weak acid, HX, in Section 5.1: the undissociated acid occurs on the left, and $H^+(aq)$ and $X^-(aq)$ on the right.

□ Write down an expression for the equilibrium constant for Equation 33.

■ Following Equation 30,

$$K = \frac{[H^+(aq)][OH^-(aq)]}{[H_2O(l)]} \qquad (34)$$

The terms $[H^+(aq)]$ and $[OH^-(aq)]$ represent the molar ion concentrations in water, but what does $[H_2O(l)]$ mean? Well, according to the measure of concentration that we have adopted, $[H_2O(l)]$ is the number of moles of water in one litre of water. One litre of water has a mass of 1 000 g, and the molar mass of water is 18 g. So, ignoring the minute amount of water

ION PRODUCT OF WATER

NEUTRAL SOLUTION

LOGARITHM TO THE BASE 10

present as ions,

$$[H_2O(l)] = \frac{1\,000}{18.0} \text{mol l}^{-1}$$
$$= 55.6 \text{mol l}^{-1}$$

In Equation 34 this constant term is *by convention* taken over to the left-hand side, so that Equation 34 becomes

$$K[H_2O(l)] = [H^+(aq)][OH^-(aq)]$$

The combination $K[H_2O(l)]$ is called the **ion product of water** and is written K_w. The experimentally determined value of K_w is $1.0 \times 10^{-14} \text{mol}^2 \text{l}^{-2}$ at 25 °C. Thus

$$K_w = [H^+(aq)][OH^-(aq)] = 1.0 \times 10^{-14} \text{mol}^2 \text{l}^{-2} \quad (35)$$

Notice, first of all, that the size of this constant does indeed limit the concentrations of hydrogen and hydroxide ions that can coexist in pure water. In water, the $H^+(aq)$ and $OH^-(aq)$ ions are supplied by the dissociation of water: every molecule that dissociates forms one hydrogen and one hydroxide ion, so if $[H^+(aq)]$ is x, $[OH^-(aq)]$ must be x as well. From Equation 35,

$$K_w = [H^+(aq)][OH^-(aq)] = x^2 = 1.0 \times 10^{-14} \text{mol}^2 \text{l}^{-2}$$

so

$$x = 1.0 \times 10^{-7} \text{mol l}^{-1}$$

This result shows that in pure water at 25 °C, the hydrogen and hydroxide ion concentrations are both $1.0 \times 10^{-7} \text{mol l}^{-1}$—very small values indeed! Roughly speaking, they mean that only one in every thousand million water molecules is ionized—Reaction 32 proceeds to only a tiny extent.

However, the implications of Equation 35 are rather more wide ranging than this. You have seen that an acid solution is one in which $[H^+(aq)]$ is high, whereas a basic solution is one in which $[OH^-(aq)]$ is high. Equation 35 expresses this idea in a quantitative way: it imposes a limitation on the concentrations of hydrogen and hydroxide ions that can coexist in *any* aqueous solution, not simply in pure water.

Consider, for example, an aqueous solution of hydrochloric acid of concentration 0.1mol l^{-1}. From Section 5.1, we know that HCl is a strong acid, which is completely dissociated into ions. Thus, the concentration of $H^+(aq)$ ions in the solution is 0.1mol l^{-1}.

□ What is the concentration of $OH^-(aq)$ ions in this solution?

■ It is $1.0 \times 10^{-13} \text{mol l}^{-1}$. This result is arrived at as follows:

$$[H^+(aq)] = 0.1 \text{mol l}^{-1} = 1.0 \times 10^{-1} \text{mol l}^{-1}$$

From Equation 35,

$$K_w = [H^+(aq)][OH^-(aq)] = 1.0 \times 10^{-14} \text{mol}^2 \text{l}^{-2} \quad (35)^*$$

$$[OH^-(aq)] = \frac{1.0 \times 10^{-14} \text{mol}^2 \text{l}^{-2}}{1.0 \times 10^{-1} \text{mol l}^{-1}}$$
$$= 1.0 \times 10^{-13} \text{mol l}^{-1}$$

□ What is the concentration of $H^+(aq)$ ions in an aqueous solution of NaOH of concentration 0.1mol l^{-1}?

■ Again, it is $1.0 \times 10^{-13} \text{mol l}^{-1}$. NaOH may also be regarded as completely dissociated in water, so $[OH^-(aq)]$ is 0.1mol l^{-1}. From Equation 35, $[H^+(aq)]$ must be $1.0 \times 10^{-13} \text{mol l}^{-1}$.

Notice that the hydrogen ion concentration in one aqueous solution may be as much as 10^{12} times greater than that in another, but it always has a finite value, even in strongly basic solutions, Similarly, the hydroxide ion concentration always has a finite value, even in strongly acidic solutions.

☐ How would you define a neutral solution?

■ A **neutral solution** shows neither acidic nor basic properties. This suggests $[H^+(aq)] = [OH^-(aq)]$ as a definition. The value of K_w suggests further that at 25 °C, $[H^+(aq)] = [OH^-(aq)] = 1.0 \times 10^{-7}\,mol\,l^{-1}$. Thus at 25 °C an acidic solution can be defined as one in which $[H^+(aq)]$ is greater than $10^{-7}\,mol\,l^{-1}$, and a basic solution is one in which it is less.

ITQ 9 A little solid calcium hydroxide is dissolved in a litre of water. Assuming that the temperature remains at 25 °C throughout, which three of the statements (a)–(e) are correct?

(a) $[OH^-(aq)]$ increases.

(b) $[H^+(aq)]$ increases.

(c) K_w increases.

(d) The number of undissociated water molecules increases.

(e) $[H^+(aq)]$ decreases.

6.2 THE pH SCALE

Try the following thought experiment. Suppose you have one litre of hydrochloric acid solution of concentration $0.1\,mol\,l^{-1}$: the concentrations of $H^+(aq)$ and $OH^-(aq)$ will be those calculated in Section 6.1, $[H^+(aq)] = 0.1\,mol\,l^{-1}$ and $[OH^-(aq)] = 1.0 \times 10^{-13}\,mol\,l^{-1}$. Suppose that you now slowly add 0.1 mol of solid sodium hydroxide. A neutralization reaction occurs: $H^+(aq)$ is consumed and the acidity of the solution drops. When all of the solid has been added, you end up with a solution of sodium chloride in water. Now the concentrations of $H^+(aq)$ and $OH^-(aq)$ will be those for pure water; that is, $[H^+(aq)] = [OH^-(aq)] = 1.0 \times 10^{-7}\,mol\,l^{-1}$. Thus, on adding the NaOH to the HCl solution, the concentration of $H^+(aq)$ decreases by a factor of 10^6 (a million) from 10^{-1} to 10^{-7}; at the same time, the hydroxide ion concentration rises by the same factor. What enormous concentration changes are associated with this simple experiment!

Suppose that you wanted to show graphically how the acidity, that is the concentration of $H^+(aq)$, changes during the course of this experiment: how do you even choose a scale for such a graph, when the quantity of interest changes by *six* orders of magnitude? Fortunately, there is a way out. As the concentration of $H^+(aq)$ drops a million-fold from 10^{-1} to 10^{-7}, this notation shows that the *index* simply drops from -1 to -7. It looks as if this index might provide a much more manageable measure of concentration and concentration changes when they cover many orders of magnitude. This index of ten is called the **logarithm to the base ten** or the common logarithm of the number in question: it is written as $\log_{10}$, or simply as log.

If you have met logarithms before, try the following ITQ to check your understanding, then skip the box that follows. If not, or if you get any of the answers wrong, read on.

ITQ 10 (a) Write down, without using tables or a calculator, the logarithms to the base ten of the following numbers: 10, 100, 0.001.

(b) Your calculator should have a special key for calculating logarithms to the base ten. Use it to calculate the logarithms of the following numbers: 4, 40, 0.4.

(c) In human blood, $[H^+(aq)] = 4.0 \times 10^{-8}\,mol\,l^{-1}$. Calculate

$$\log\left\{\frac{[H^+(aq)]}{mol\,l^{-1}}\right\},$$

that is, $\log(4.0 \times 10^{-8})$.

pH SCALE

WORKING WITH LOGARITHMS

You have seen, in **MAFS 1**, that 10 can be written in scientific notation as 10^1. Then according to the definition of a logarithm,

$$\log 10 = \log 10^1 = 1$$

Similarly, $\log 100 = \log 10^2 = 2$

and $\log 1\,000 = \log 10^3 = 3$

You could, of course, continue with this process, but these three examples are sufficient to illustrate an extremely important property of logarithms.

Suppose you start by writing 1 000 as (10×100) instead of $(10 \times 10 \times 10)$. Then

$$\log 1\,000 = \log (10 \times 100) = \log (10^1 \times 10^2)$$

but $\log 1\,000 = 3$

So $\log (10^1 \times 10^2) = 3 \qquad (36)$

However, $\log 10^1 = 1$ and $\log 10^2 = 2$

So $\log 10^1 + \log 10^2 = 3 \qquad (37)$

Comparing Equations 36 and 37, you can see that

$$\log (10^1 \times 10^2) = \log 10^1 + \log 10^2$$

This is, in fact, a completely general result: *the logarithm of two numbers multiplied together is equal to the sum of the logarithms of each number*, that is

$$\log (x \times y) = \log x + \log y \qquad (38)$$

Now, according to the definition of log,

$$\log 0.1 = \log 10^{-1} = -1$$

and

$$\log 0.01 = \log 10^{-2} = -2, \text{ etc.}$$

But what is the logarithm to the base 10 of unity? Well, according to Equation 38, you can write

$$\begin{aligned} \log 1 &= \log (0.1 \times 10) \\ &= \log 10^{-1} + \log 10^1 \\ &= -1 + 1 \\ &= 0 \end{aligned}$$

In other words, 1 can be written in index notion as 10^0.

So, for any number (including unity) that can be expressed as a positive or negative whole number power of ten, the common logarithm can be written down by inspection. This is a skill you must master: for example, you may be required to use it during the examination when calculators are prohibited. The skill is tested in ITQ 11.

You should now know nearly as much about logarithms as you will need for the moment. There remains just one further problem: how do you obtain the logarithm of a number that cannot be expressed as a simple whole number power of ten? Suppose, for example, that you have a solution containing $[H^+(aq)] = 4.0\,mol\,l^{-1}$; what is the value of

$$\log \left\{ \frac{[H^+(aq)]}{mol\,l^{-1}} \right\}$$

in this case? What you need to do is express 4.0 in the form 10^x, but how do you determine the value of x? This time you will need your calculator to do it for you. Check that you know how to use it by going back and working through the examples in ITQ 10.

Table 4 shows values of

$$\log\left\{\frac{[H^+(aq)]}{mol\,l^{-1}}\right\}$$

for hydrogen ion concentrations within the range of interest of most chemists: do not worry about the column labelled pH for the moment.

TABLE 4 Values of $[H^+(aq)]$, with corresponding values of $\log\left\{\frac{[H^+(aq)]}{mol\,l^{-1}}\right\}$ and pH (at 25 °C)

$[H^+(aq)]/mol\,l^{-1}$	$\log\left\{\frac{[H^+(aq)]}{mol\,l^{-1}}\right\}$	pH	
10^{1} or 10	1	−1	↑
10^{0} or 1	0	0	
10^{-1} or 0.1	−1	+1	increasingly
10^{-2} or 0.01	−2	+2	acidic
10^{-3} or 0.001	−3	+3	
10^{-4} or 0.000 1	−4	+4	
10^{-5} or 0.000 01	−5	+5	
10^{-6} or 0.000 001	−6	+6	pure water
10^{-7} or 0.000 000 1	−7	+7	neutral
10^{-8} or 0.000 000 01	−8	+7.4 blood	
10^{-9} or 0.000 000 001	−9	+8	
10^{-10} or 0.000 000 000 1	−10	etc	increasingly
10^{-11} or 0.000 000 000 01	−11		basic
10^{-12} or 0.000 000 000 001	−12		
10^{-13} or 0.000 000 000 000 1	−13		
10^{-14} or 0.000 000 000 000 01	−14		↓

□ In ITQ 10 you found that log 4 = 0.602 1. Does this value seem reasonable when compared with the values of log 1 and log 10 in Table 4?

■ Well, 4 is a number between 1 and 10, so according to Table 4, log 4 should lie between 0 and 1: it does.

But the values in Table 4 indicate that for the concentration range of interest the logarithm itself will almost invariably be a negative number. For this reason, a widely used scale is defined in terms of the *negative* of the logarithm. This arbitrary, but useful, scale is called the **pH scale**:

$$pH = -\log\left\{\frac{[H^+(aq)]}{mol\,l^{-1}}\right\} \tag{39}$$

□ Fill in the missing pH values in Table 4. What are the pH values of pure water and of human blood at 25 °C ($[H^+(aq)] = 4.0 \times 10^{-8}\,mol\,l^{-1}$, as given in ITQ 10)?

■ For pure water $[H^+(aq)] = 10^{-7}\,mol\,l^{-1}$, so

$$\log\left\{\frac{[H^+(aq)]}{mol\,l^{-1}}\right\} = -7.0 \text{ and the pH} = +7.0.$$

For human blood,

$$\log\left\{\frac{[H^+(aq)]}{mol\,l^{-1}}\right\} = \log(4.0 \times 10^{-8})$$

$$= \log(4.0) + \log(10^{-8})$$

$$= (0.6021 - 8.0000) = -7.3979.$$

So pH = +7.4 (to two significant figures).

Further examples of this type of calculation are provided in ITQ 11 and in the SAQs at the end of Section 6. For the moment, try to familiarize yourself with the pH scale by studying Table 4. Marked against the figures are the acidic and basic regions of the scale. You will remember from Section 6.1 that acidic solutions have $[H^+(aq)]$ greater than 10^{-7} mol l^{-1}, so their pH is *less* than 7. Conversely, basic solutions have pH greater than 7. To give you a better feel for pH, values of pH for some common chemical, natural, and household solutions are given in Table 5.

TABLE 5 pH values for some common solutions

Solution	pH
hydrochloric acid (0.1 mol l^{-1})	1.0
gastric juice (human)	1.0–2.5
lemon juice	about 2.1
acetic acid (0.1 mol l^{-1})*	2.9
orange juice	about 3.0
tomato juice	about 4.1
urine	6.0
rainwater (unpolluted)	5.2–6.5
saliva (human)	6.8
milk	about 6.9
pure water (25 °C)	7.0
blood (human)	7.4
seawater	7.9–8.3
ammonia (0.1 mol l^{-1})†	11.1
sodium hydroxide (0.1 mol l^{-1})	13.0

* Approximately the concentration of household vinegar.
† Approximately the concentration of household ammonia solutions.

There are instruments called pH meters, which can be used to measure the pH of a given solution directly: many of the values in Table 5 were obtained in this way.

ITQ 11 Consider the numbers 1, 0.1, 0.01 and 0.000 001.

(i) Write each of them in the form 10^n, where n is an integer.

(ii) Write down the logarithm to the base 10 of each number.

(iii) What is the pH of the solutions in which $[H^+(aq)]$ is 1 mol l^{-1}, 0.1 mol l^{-1} and 0.000 001 mol l^{-1}, respectively?

ITQ 12 Look at the entries in Table 5. Which is more acidic: orange juice or tomato juice? Vinegar or gastric juice? Blood or seawater?

6.3 ACID RAIN

In Experiment 1, you showed that carbon dioxide, a gas that you breathe out, dissolves in water to give a slightly acid solution:

□ What reaction is responsible for the acidity?

■ From Section 2.4, it is the equilibrium

$$CO_2(g) + H_2O(l) \rightleftharpoons HCO_3^-(aq) + H^+(aq) \quad (40)$$

Carbon dioxide comprises about 0.03% of the Earth's atmosphere. Table 5 shows that unpolluted rainwater has a pH in the range 5.2–6.5. This slight acidity is explained by dissolved carbon dioxide that has undergone the reaction in Equation 40.

The acidity of CO_2 solutions can be set in a more general context. Two points are worth making. First, CO_2 is the oxide of a non-metallic element, carbon, and *if they are significantly soluble in water, the oxides of non-metals always give acid solutions.* By contrast, the oxides of metals may be either bases (MgO in SAQ 4) or acids (CrO_3 whose acid solution you will study at Summer School). Secondly, an element sometimes forms more than one normal oxide. Where this is so, *the higher oxide is always the more acidic.*

Rainwater is only slightly acid because Equilibrium 40 lies well over to the left: CO_2 is a weak acid. But modern industrial economies generate other non-metallic oxides, which can increase the acidity of rainwater if they are discharged into the atmosphere. The problem can be traced back to two chemical culprits. First, sulphur compounds in coal and oil are converted into sulphur dioxide gas, SO_2, when the fuel is burnt in power stations. Secondly, spark temperatures in a car engine can exceed 2 000 °C, and at these high temperatures, some atmospheric nitrogen and oxygen can combine to form gaseous nitric oxide, NO:

$$N_2(g) + O_2(g) = 2NO(g) \tag{41}$$

Up to 0.4% of the exhaust gases from an accelerating motor car can consist of nitric oxide. Substantial amounts are also produced in power stations.

The acidity of NO is negligible, and that of SO_2 is only weak. But both compounds can react with other atmospheric gases to form higher oxides of nitrogen and sulphur, whose solutions are therefore much more acidic. The details of these reactions are complicated. There are a number of steps, and other chemicals such as ozone and certain peroxides are produced and consumed. However, the *net results* are fairly easy to describe. Sulphur dioxide takes on atmospheric oxygen to give the trioxide:

$$2SO_2(g) + O_2(g) = 2SO_3(g) \tag{42}$$

SO_3 reacts with water vapour to give sulphuric acid, H_2SO_4, a strong acid like HCl and HNO_3. When dissolved in rainwater, H_2SO_4 is completely dissociated, principally into $H^+(aq)$ and sulphate ions, $SO_4{}^{2-}(aq)$:

$$H_2SO_4(aq) = 2H^+(aq) + SO_4{}^{2-}(aq) \tag{5*}$$

The overall reaction is, therefore,

$$2SO_2(g) + O_2(g) + 2H_2O(l) = 4H^+(aq) + 2SO_4{}^{2-}(aq) \tag{43}$$

In a similar fashion, nitric oxide can be converted into nitric acid, which then dissolves in rainwater. The overall reaction is

$$4NO(g) + 3O_2(g) + 2H_2O(l) = 4H^+(aq) + 4NO_3{}^-(aq) \tag{44}$$

Since 1950, SO_2 and NO emissions in Europe have approximately doubled, and the acidity of rainwater has increased. On 10 April 1974, a rainstorm at Pitlochry in Scotland had a pH of 2.4, an acidity greater than that of vinegar, but that was exceptional. More typical are the data in Table 6, which compare rainwater compositions from similar sites in inland Scandinavia during the 1950s and 1970s. Data from a polluted site in the United States are also included.

TABLE 6 Ion concentrations and pH of rainwater from Scandinavian and US sites in 1956 and 1974

Site	pH	$[H^+(aq)]$ / 10^{-6} mol l^{-1}	$[SO_4{}^{2-}(aq)]$ / 10^{-6} mol l^{-1}	$[NO_3{}^-(aq)]$ / 10^{-6} mol l^{-1}	$[HCO_3{}^-(aq)]$ / 10^{-6} mol l^{-1}
inland Scandinavia (1956)	5.4	4	15	0	6
inland Scandinavia (1974)	4.3	48	26	26	0
inland north-eastern United States (1974)	3.9	114	55	50	0

ACID RAIN

□ How do these data corroborate our explanation of the origins of acid rain?

■ Higher acidity and lower pH are associated with high concentrations of sulphate and nitrate, the ions produced alongside $H^+(aq)$ in Equations 43 and 44.

Figure 7 shows average pH contours for western European rain between 1978 and 1982. The low of 4.1 hovers over central Europe.

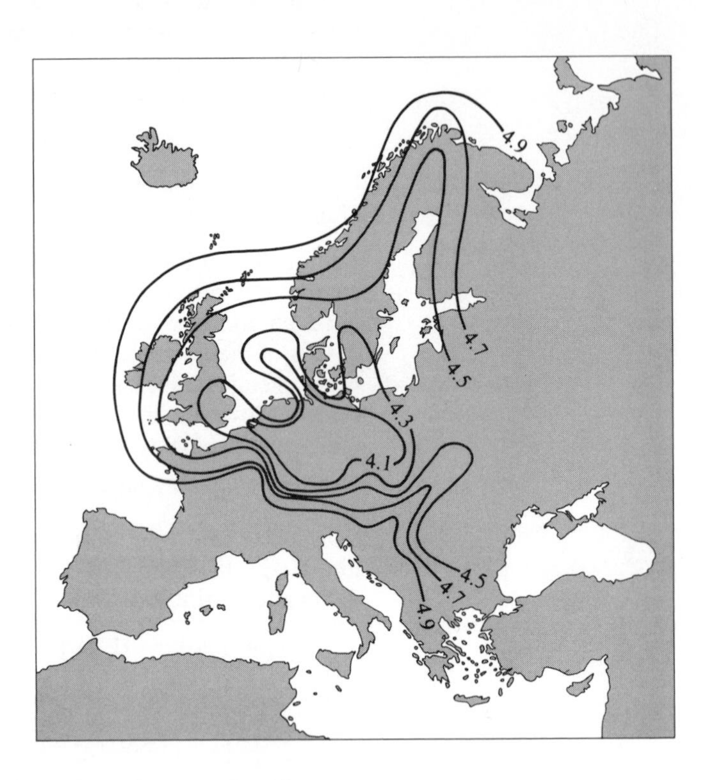

FIGURE 7 Mean pH values for rainfall over western Europe in the period 1978–82.

Acid rain is a political issue because those regions responsible for it are not always the chief sufferers. Tall stacks on power stations (Figure 8) may well export the problem sometimes as far as other countries. In Europe, the prevailing winds have a northerly drift and converge on Scandinavia. Thus, it has been estimated that only 8% of the sulphate falling on Norway is due to the activities of Norwegians, and that 17% comes from the UK. By contrast 79% of UK sulphate deposition is of British origin, and none is Norwegian.

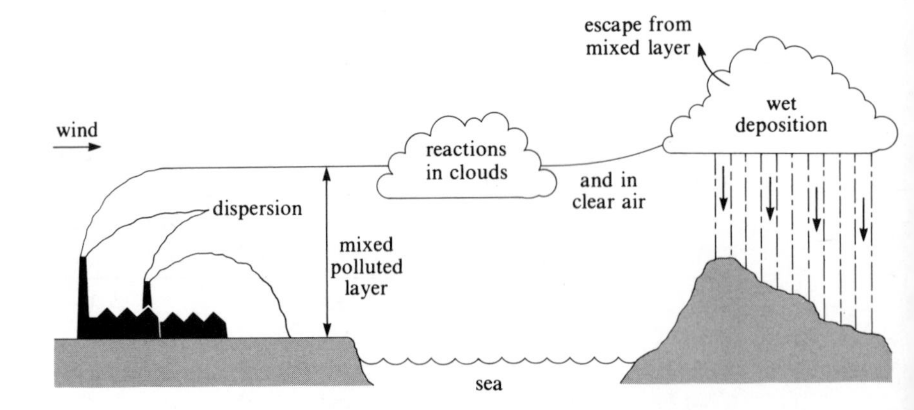

FIGURE 8 Dispersion of pollutants such as SO_2 and NO leading subsequently to acid rain.

The vulnerability of different regions also varies widely. What is important is the capacity of the environment to neutralize the acidity after the rain touches ground. Land containing forms of calcium carbonate, such as chalk or limestone, is effective in this role because of the reaction with acid that is quoted in the answer to ITQ 1:

$$CaCO_3(s) + 2H^+(aq) = Ca^{2+}(aq) + CO_2(g) + H_2O(l) \qquad (45)$$

Unfortunately, the bedrock of Scandinavia consists mainly of rocks, such as granite or quartz sandstone, that are poor neutralizers of acidity. In such an environment, the impact of acid rain is likely to be more severe.

During the past 50 years, trout and salmon have disappeared from some 2 000 lakes in southern Scandinavia. The influence of acid rain on something as complicated as the countryside is very difficult to assess, and wild and unsubstantiated claims have been made. However, there is strong evidence that acid rain is responsible for this particular catastrophe. Norway and Sweden have called for a reduction of SO_2 and NO emissions throughout Europe. In 1986, the UK government announced projected expenditure of £600 million on chemical plant to cut down the SO_2 in power station emissions, and to convert the gas into a marketable product, which will be calcium sulphate, sulphuric acid or sulphur. This action was judged inadequate by the Norwegian government.

SUMMARY OF SECTION 6

1 Pure water contains tiny amounts of aqueous hydrogen and hydroxide ions.

2 The ion product of water

$$K_w = [H^+(aq)][OH^-(aq)] = 1.0 \times 10^{-14}\,\text{mol}^2\,\text{l}^{-2} \text{ at } 25\,°\text{C}$$

controls the concentrations of hydrogen and hydroxide ions that can coexist in *any* aqueous solution. If either $[H^+(aq)]$ or $[OH^-(aq)]$ is known, the constant value of K_w allows the other to be calculated.

3 The concentration of aqueous hydrogen ions in a solution can be expressed by stating its pH, which is defined as

$$\text{pH} = -\log\left\{\frac{[H^+(aq)]}{\text{mol}\,\text{l}^{-1}}\right\}$$

4 The combustion of fossil fuels in power stations and motor vehicles produces the oxides SO_2 and NO. These react with atmospheric oxygen to form more acidic substances, such as sulphuric and nitric acid, which dissolve in rainwater and yield acid rain.

SAQ 13 In Section 6.1, we said that in pure water at 25 °C, the ratio of ionized to un-ionized water molecules is about one to every thousand million. Confirm this by calculating a more accurate value from the concentrations worked out in Section 6.1.

SAQ 14 0.001 mole of nitric acid is dissolved in 10 litres of water at 25 °C. Calculate the hydrogen ion concentration, $[H^+(aq)]$, the pH, and the hydroxide ion concentration, $[OH^-(aq)]$, of the resulting solution.

SAQ 15 The solubility of $Ca(OH)_2$ in water is $0.020\,\text{mol}\,\text{l}^{-1}$ at 25 °C. Calculate

(i) The value of $[OH^-(aq)]$ in the saturated solution of $Ca(OH)_2$ at 25 °C;

(ii) The corresponding value of $[H^+(aq)]$, given that K_w, the ion product of water, is $1.0 \times 10^{-14}\,\text{mol}^2\,\text{l}^{-2}$;

(iii) The pH of the saturated solution at 25 °C.

SAQ 16 Chlorine forms the oxides Cl_2O and Cl_2O_7 which, on reaction with water, yield the acids HClO and $HClO_4$, respectively. One of these acids is strong, like nitric acid, and the other is weak, even weaker than acetic acid.

(i) By inspecting the formulae of the oxides, decide which one reacts with water to form the strong acid. Write an equation for this reaction with water.

(ii) Which acid is the weak one of the two? Write an equation for its dissociation in aqueous solution.

SAQ 17 0.0572 mole of pure acetic acid is dissolved in a little water, and more water is then added until the total volume of the solution is one litre. The pH of the solution is then measured with a pH meter, and found to be 3.00.

(i) Calculate a value of the acid dissociation constant K_a, and compare it with the value of Table 3.

(ii) What is the ratio of un-ionized to ionized acetic acid molecules in the solution?

7 GENERALIZED FORM OF THE EQUILIBRIUM CONSTANT

Any chemical equilibrium at constant temperature has an equilibrium constant that provides a precise numerical relationship between the concentrations of the ingredients of the equilibrium system. In Sections 5 to 6.2, we illustrated this with equilibria of the general type

$$HX(aq) \rightleftharpoons H^+(aq) + X^-(aq) \qquad (29)^*$$

for which the equilibrium constant is

$$K = \frac{[H^+(aq)][X^-(aq)]}{[HX(aq)]} \qquad (30)^*$$

These illustrative examples were of a very particular kind. First, all species in the equilibria are in aqueous solution. Secondly, the numbers preceding the formulae in the equations are all equal to one. Thus, *one* HX(aq) molecule breaks up into *one* H^+(aq) ion and *one* X^-(aq) ion. So what form does an equilibrium constant take when we break free of these narrow restrictions?

Consider one of the examples from Section 1: a mixture of hydrogen and iodine is heated in a sealed vessel at 427 °C. The purple colour of the iodine vapour decreases in intensity as the hydrogen and iodine react to form a colourless gas, hydrogen iodide, HI:

$$H_2(g) + I_2(g) = 2HI(g) \qquad (2)^*$$

Eventually, there is no further change in the intensity of the colour: equilibrium has been reached. Although hydrogen and iodine molecules are still reacting together to form hydrogen iodide, this forward reaction is exactly balanced by hydrogen iodide molecules decomposing. We can now describe the system with the equilibrium equation

$$H_2(g) + I_2(g) \rightleftharpoons 2HI(g) \qquad (46)$$

Notice how this example violates both of our previous restrictions. First, one formula in the equilibrium equation, that of HI, is preceded by a 2. Secondly, all the species in the equilibrium are gases. Note, however, that at equilibrium some definite number of moles of each gas will be present, and that the bulb will have a definite volume of, say, V litres. Each gas will be evenly dispersed throughout the bulb, so we can assign to it an equilibrium concentration in mol l^{-1}, which is obtained by dividing the number of moles by V.

What is the numerical relationship between these equilibrium concentrations? It is supplied by the equilibrium constant, K, of Equation 46, which takes the form

$$K = \frac{[HI(g)]^2}{[H_2(g)][I_2(g)]} \qquad (47)$$

At 427 °C, experiments show that K is close to 54.

Look carefully at Equation 47. Unlike Equation 30, it refers to a chemical equation, Equation 46, in which one of the formulae is preceded by a 2.

□ How does this 2 display its presence in Equations 46 and 47?

■ In Equation 46, it precedes the formula of HI(g). In Equation 47, *it becomes a power to which the concentration of* HI(g) *is raised.*

To reinforce this point, recall our observation in Section 5: we noted that the equilibrium constant for Equation 25 could be obtained by multiplying the concentrations of the chemicals on the right-hand side together, and then dividing by the product of the concentrations of the chemicals on the left-hand side. If equation 46 is written,

$$H_2(g) + I_2(g) = HI(g) + HI(g) \tag{48}$$

this same procedure yields Equation 47.

With these insights, you can see that Equations 30 and 47 are particular examples of a general case. Suppose reactants A, B, . . . etc. yield products P, Q, . . . etc. in a reaction that reaches equilibrium:

$$a\text{A} + b\text{B} + \ldots \rightleftharpoons p\text{P} + q\text{Q} + \ldots \tag{49}$$

Here, a, b, p, q are the numbers used to balance the equation. Then the equilibrium constant takes the form,

$$K = \frac{[\text{P}]^p[\text{Q}]^q \ldots}{[\text{A}]^a[\text{B}]^b \ldots} \tag{50}$$

Notice that the concentration of each ingredient in Equation 50 is raised to a power equal to the number that precedes the ingredient in Equation 49. Thus adjusted, the concentrations of the products are multiplied together in the numerator of the equilibrium constant; those of the reactants are multiplied in the denominator.

During this procedure, the chemical Equation 49 fathers the equilibrium constant in Equation 50: until the first is specified, the second cannot be written down. *Any expression for an equilibrium constant is inextricably tied to a particular chemical equation from which it is derived.*

You will find an example that emphasizes this point in parts (c) and (d) of the following ITQ.

ITQ 13 Write down expressions for the equilibrium constants of the following reactions:

(a) $N_2(g) + O_2(g) \rightleftharpoons 2NO(g)$

(b) $Sn^{2+}(aq) + 2Fe^{3+}(aq) \rightleftharpoons Sn^{4+}(aq) + 2Fe^{2+}(aq)$

(c) $2H_2(g) + O_2(g) \rightleftharpoons 2H_2O(g)$

(d) $H_2(g) + \frac{1}{2}O_2(g) \rightleftharpoons H_2O(g)$

Because product concentrations appear in the numerators of equilibrium constants, and reactant concentrations in the denominators, K is large when, at equilibrium, the products in the parent equation are plentiful, and reactants are scarce. For this reason, K is a measure of the extent of reaction—of how far the reaction can proceed in a left to right direction when pure reactants are mixed and maintained at the temperature in question. Two extreme cases are supplied by two reactions in ITQ 13:

$$2H_2(g) + O_2(g) \rightleftharpoons 2H_2O(g); \qquad K = 3.3 \times 10^{81}\,\text{l mol}^{-1} \text{ at } 25\,°\text{C}$$

$$N_2(g) + O_2(g) \rightleftharpoons 2NO(g); \qquad K = 7.2 \times 10^{-31} \text{ at } 25\,°\text{C}$$

For the first reaction, K is very large: evidently equilibrium lies far over to the right-hand side, and the reaction should be effectively complete when equilibrium is attained. By contrast, K for the second reaction is tiny: you would expect a vanishingly small amount of product to be present at equilibrium. This point was made in Section 5.1 with reference to a specific type of reaction: the dissociation of an acid. Here it has been applied to reactions in general. This generalization of things learnt in Section 5 is also an important feature of SAQs 18 to 20, which you should now try.

SUMMARY OF SECTION 7

1 The equilibrium constant of the general equilibrium,

$$aA + bB + \ldots \rightleftharpoons pP + qQ + \ldots \qquad (49)^*$$

is given by

$$K = \frac{[P]^p[Q]^q \ldots}{[A]^a[B]^b \ldots} \qquad (50)^*$$

Thus the expression for the equilibrium constant depends crucially on the form of the balanced equation that is used to describe the equilibrium.

2 The size of the equilibrium constant at a particular temperature is a measure of how far the reaction will have gone when equilibrium is attained at that temperature. In other words, it indicates the equilibrium yield of the reaction at that temperature.

SAQ 18 In Section 7, you were told that, at 427 °C, the equilibrium constant of the reaction

$$H_2(g) + I_2(g) \rightleftharpoons 2HI(g)$$

is about 54. In Section 1, you were told of an equilibrium mixture for this system which contained 0.35 mole of $H_2(g)$, 0.15 mole of $I_2(g)$ and 1.70 moles of $HI(g)$ in a sealed bulb at 427 °C. Show that this composition is consistent with the equilibrium constant. Assume that the volume of the bulb is 1 litre.

SAQ 19 Write down an expression for the equilibrium constant of the reaction mentioned in SAQ 8, namely

$$N_2(g) + 3H_2(g) \rightleftharpoons 2NH_3(g)$$

Use this expression to predict the effect of removing NH_3 from an equilibrium mixture by dissolving it in water. How does this compare with the predictions derived from Le Chatelier's principle?

SAQ 20 At 25 °C, the equilibrium constant, K, of the reaction,

$$N_2(g) + 3H_2(g) \rightleftharpoons 2NH_3(g)$$

is $2.70 \times 10^8\,\text{mol}^{-2}\,\text{l}^2$. A steel bulb of volume 1 litre contains 0.0100 mole of $H_2(g)$ and 0.0033 mole of $N_2(g)$ at 25 °C. Assuming that the gases are then in equilibrium with ammonia:

(i) How many moles of ammonia does the bulb contain?

(ii) If the equilibrium state was set up by pumping $N_2(g)$ and $H_2(g)$ into the empty bulb, and then allowing them to reach equilibrium, how many moles of $N_2(g)$ and of $H_2(g)$ were pumped in?

8 TV NOTES: EQUILIBRIUM RULES—OK?

The key points in the programme are summarized below.

1 A pH meter (an instrument that you will use at Summer School) was used to show that a typical weak acid, acetic acid (HAc), is only partly dissociated into ions in solution; a dynamic equilibrium is set up:

$$\text{HAc(aq)} \rightleftharpoons \text{H}^+\text{(aq)} + \text{Ac}^-\text{(aq)} \qquad (25)^*$$

2 The idea of a dynamic equilibrium was examined with the help of a simple analogy—our 'equilibrium game'. If you would like to try it at home, you will need the equipment illustrated in Figure 9. The 'rules of the game' simply involve deciding how many squares the 'marker' on each side is allowed to move in any one turn: we chose two for the left-hand player and three for the right-hand one. Why not try some different values and see if you can predict what will happen?

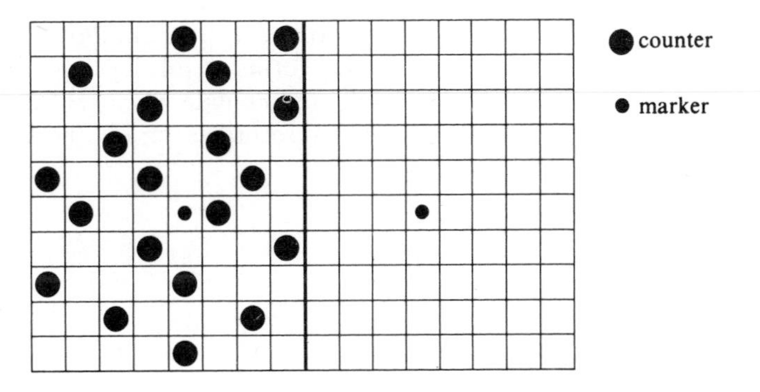

FIGURE 9 Equipment for the 'equilibrium game'. The size of the board is not critical, but it should be divided into two equal parts. We used 20 counters.

The main object of the game was to show that it is possible to build up and maintain a roughly constant distribution of counters between the two players, even though individual counters are still going back and forth. But the analogy should not be taken too far. Strictly speaking, the game should only be compared with a simple equilibrium between two species:

$$\underset{\text{(left)}}{\text{A}} \rightleftharpoons \underset{\text{(right)}}{\text{B}} \qquad (51)$$

Even here, there are problems. The reactant and product in a chemical system are not in separate compartments; neither is there a marker that 'magically' converts one into the other!

Despite these reservations, you saw that our artificial equilibrium responded to a disturbance (removal of all the counters from the right-hand side) in much the same way as a chemical system: there was a net movement from left to right until a new equilibrium was established.

3 We used these ideas to interpret the observed effects of disturbing two chemical systems. Adding acetate ions to an acetic acid solution shifted the equilibrium in Equation 25 to the left, as shown by a marked fall in the concentration of H^+(aq). Compare the exercises in ITQ 7.

We then showed that Ian Soutar's splendid trick revolves around the manipulation of the following simple equilibrium:

$$\underset{\text{colourless}}{\text{HIn(aq)}} \rightleftharpoons \text{H}^+\text{(aq)} + \underset{\text{red}}{\text{In}^-\text{(aq)}} \qquad (52)$$

where HIn represents an indicator (phenolphthalein). The equilibrium was shifted back and forth (and hence the colour of the solution changed) by adjusting the concentration of H^+(aq).

These effects are summarized by Le Chatelier's principle.

4 The final part of the programme introduced an experimental technique that you will use at Summer School—absorption spectrophotometry. You saw how this technique can, in principle, be used to determine the equilibrium constant K for a second indicator, bromophenol blue:

$$\underset{\text{yellow}}{HIn(aq)} \rightleftharpoons H^+(aq) + \underset{\text{purple}}{In^-(aq)} \qquad (53)$$

where

$$K = \frac{[H^+(aq)][In^-(aq)]}{[HIn(aq)]}$$

As you saw, the technique depends on the fact that both forms of the indicator (HIn and In^-) absorb light, but in different regions of the visible spectrum (Figure 10). Thus, for example, the purple solution of In^- absorbs light mainly between 500 nm and 620 nm—the green, yellow and orange region. The transmitted light, which is what we see, corresponds to a mixture of the remaining blues and red.

When will the purple form predominate?

According to the equilibrium in Equation 53, the balance between the coloured forms again depends on the concentration of $H^+(aq)$: thus, $In^-(aq)$ will predominate when $[H^+(aq)]$ is low, and vice versa. This is how we were able to record spectra of the two 'pure' species (Figure 10).

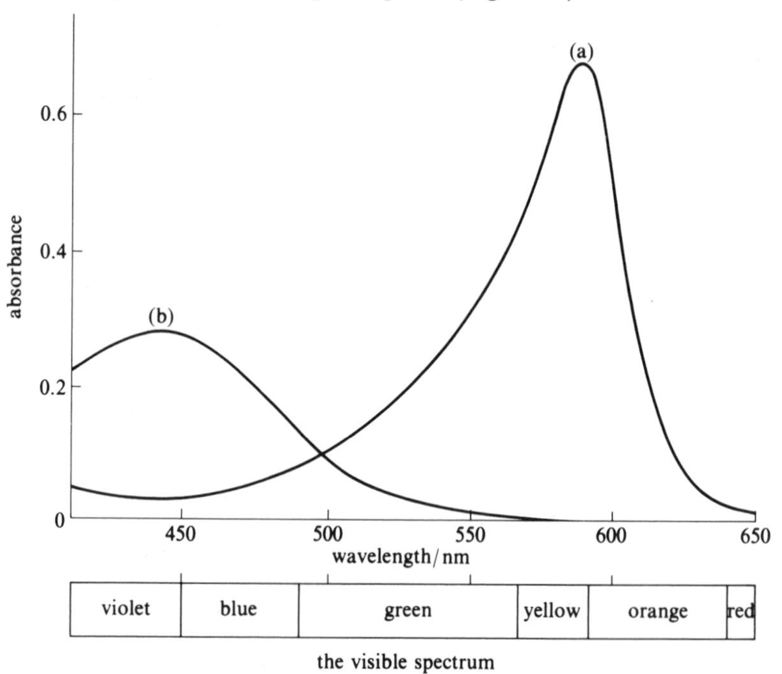

FIGURE 10 Absorption spectra of bromophenol blue related to the visible spectrum: (a) when $[H^+(aq)]$ is low, $In^-(aq)$ predominates; (b) when $[H^+(aq)]$ is high, HIn(aq) predominates.

You will see at Summer School how we can use such spectra to calculate the relative equilibrium concentrations of the two forms in a solution of intermediate pH that contains them both.

The values we obtained are given in Table 7, where the hydrogen ion concentration is expressed as the pH, as measured by a pH meter. When you have read about pH in Section 6.2, you should complete this Table and thus convince yourself that K is indeed close to a constant.

TABLE 7 Experimental results for the dissociation of bromophenol blue

pH	$[H^+(aq)]$/ $mol\,l^{-1}$	$[In^-(aq)]$/ $[HIn(aq)]$	$K/mol\,l^{-1}$
4.6		3.80	
4.3	5.01×10^{-5}	1.77	8.87×10^{-5}
3.6		0.33	

5 Finally, you saw that the equilibrium in Equation 54 can be shifted over to the right (the colour darkened) by raising the temperature

$$N_2O_4(g) \rightleftharpoons 2NO_2(g) \qquad (54)$$

dinitrogen tetroxide (colourless) — nitrogen dioxide (brown)

The equilibrium constant for this reaction is given by

$$K = \frac{[NO_2(g)]^2}{[N_2O_4(g)]}$$

If $[NO_2(g)]$ increases, $[N_2O_4(g)]$ must decrease (we did not add any material to the flask), which together imply that K must increase with increasing temperature. In other words, the value of K depends on the temperature. The reasons why this should be so are taken up in Section 6 of Unit 16.

9 UNIT SUMMARY AND A FORWARD LOOK

Two major subjects were introduced in this Unit: the concept of acids and bases, and the idea of chemical equilibrium. You saw that chemical reactions, if they occur at detectable speeds, can be said to be moving towards chemical equilibrium. In Sections 5 and 6, you were shown how an equilibrium position, at some fixed temperature, can be characterized by an equilibrium constant whose size gives us an idea of how far the reaction must go before equilibrium is reached.

The preceding paragraph contains two important qualifications. The second sentence implies that chemical equilibrium will *not* be reached if *the reaction does not proceed at a detectable speed.* In other words, extreme slowness of reaction can block the path to equilibrium. Such a problem was deliberately avoided in Unit 15. Indeed this was one reason why acids, bases, electrolytes and aqueous ions were so prominent. Studying the reactions of such substances, it is easy to be selective, and choose reactants that set off determinedly towards equilibrium when they are brought into contact! In Unit 16 however, we shall move away from such systems and this will force us to confront the question of *rates* of reaction more directly. From the problem of *how far* a reaction goes, we shall shift to the problem of *how fast* it goes as it moves towards equilibrium.

The second qualification in our first paragraph is the recognition that equilibrium constants must be defined at *some fixed temperature.* As noted in Section 4.1, this is necessary because equilibrium positions and, therefore, equilibrium constants, vary with temperature. For example, from Section 7, at 427 °C, the equilibrium constant of the reaction

$$H_2(g) + I_2(g) \rightleftharpoons 2HI(g) \qquad (46)^*$$

is 54. However, at 227 °C it is 128, and at 627 °C it is 35. In some cases therefore, such as Reaction 46, the equilibrium constant decreases with increasing temperature; in other cases, such as Reaction 54, it increases. To an industrialist, anxious to obtain as much product as possible from an equilibrium system, the question of what determines such a change of yield with temperature is of great importance. This question is also answered in Unit 16, which is concerned with chemical energetics.

OBJECTIVES FOR UNIT 15

After you have worked through this Unit, you should be able to:

1 Explain the meaning of, and use correctly, all the terms flagged in the text.

2 Recall the operational definitions of an acid and a base, and classify given substances accordingly. (*Experiment 1, ITQ 1, SAQs 1 and 4*).

3 Use the Arrhenius definitions of an acid and a base to interpret the reactions of acids and bases (*ITQs 1–3, SAQs 1–4*).

4 Calculate in moles per litre the concentration of a solution, or of the ions in a solution, given the mass of solute in a given volume of solution, appropriate relative atomic masses, and the dissolution reaction of the solute (*ITQs 4 and 5, SAQ 5*).

5 Use Le Chatelier's principle to predict the consequences of a change in the concentration of a constituent of a chemical equilibrium (*ITQs 6 and 7, SAQs 7 and 8*).

6 Given a list of strong electrolytes that are very sparingly soluble, identify pairs of strong electrolyte solutions that will produce precipitates when they are mixed (*SAQ 6*).

7 Write down the expressions for the equilibrium constants for the dissociation of a weak acid, K_a, and for the dissociation of water K_w (*ITQs 8 and 9, SAQs 10–15 and 17*).

8 Show how the consequences of change in the concentration of a constituent of a chemical equilibrium at constant temperature can be predicted from the requirement that the equilibrium constant is constant under these conditions (*ITQs 8 and 9, SAQ 19*).

9 Given appropriate concentrations, calculate unknown concentrations at equilibrium from equilibrium constants, and unknown equilibrium constants from concentrations (*SAQs 10–15, 17, 18 and 20*).

10 Distinguish the meanings of the words 'strength' and 'concentration', as applied to solutions of electrolytes (*SAQs 9, 12 and 16*).

11 Define the logarithm to the base 10 of a number and, when given a number that can be written as 10^n, where n is a positive or negative integer, state what its logarithm to the base 10 is (*ITQs 10a and 11; SAQ 14*).

12 Use a pocket calculator to find the logarithms to the base 10 of a given number or product of numbers (*ITQs 10b and 10c; SAQ 15*).

13 Recognize that oxides of non-metals are usually acidic, and that higher oxides are more acidic than lower ones (*SAQ 16*).

14 Calculate the pH of a solution from either its hydrogen ion concentration or its hydroxide ion concentration, or interpret a given value of pH (*ITQs 10 and 11; SAQs 14, 15 and 17*).

15 Write down the equilibrium constant of a reaction that occurs in aqueous solution or the gas phase, given the balanced equation for the reaction (*ITQ 13; SAQs 18–20*).

ITQ ANSWERS AND COMMENTS

ITQ 1 For the reaction of the given substances with an acid, $H^+(aq)$:

(a) $Mg(s) + 2H^+(aq) = Mg^{2+}(aq) + H_2(g)$

(b) $CaCO_3(s) + 2H^+(aq) = Ca^{2+}(aq) + CO_2(g) + H_2O(l)$

Notice that although the total charge on each side of these equations is *not* zero, the equations are nevertheless balanced: the total charge on each side of the equation is the same$(+2)$.

ITQ 2 Aqueous solutions of KOH contain the ions $K^+(aq)$ and $OH^-(aq)$; aqueous solutions of HNO_3 contain the ions $H^+(aq)$ and $NO_3^-(aq)$. By analogy with Equation 10, the reaction could be written

$$K^+(aq) + OH^-(aq) + H^+(aq) + NO_3^-(aq)$$
$$= K^+(aq) + NO_3^-(aq) + H_2O(l)$$

But $K^+(aq)$ and $NO_3^-(aq)$ occur on both sides so they can be eliminated to give

$$H^+(aq) + OH^-(aq) = H_2O(l)$$

This is just Equation 11 again: at the level of ions and molecules, the reaction between KOH and HNO_3 is identical with that between NaOH and HCl.

ITQ 3 Potassium nitrate, KNO_3, which is a salt. When the reaction of ITQ 2 is over, the aqueous solution contains $K^+(aq)$, $NO_3^-(aq)$ and water. If the water evaporates, the solid left behind contains the potassium cation of the base KOH, and the nitrate anion of the acid HNO_3.

ITQ 4 $25\,g\,l^{-1}$. Every $200\,cm^3$ of solution contains 5 g of sugar, so a litre ($1\,000\,cm^3$) of such a solution will contain 25 g of sugar.

ITQ 5 With the molecular formula $C_{12}H_{22}O_{11}$, one mole of sucrose has a mass (in grams) of

$$[(12 \times 12) + (22 \times 1) + (11 \times 16)]\,g = 342\,g$$

Each litre of solution contains 25 g which is

$$\frac{25}{342}\,mol = 0.0731\,mol$$

Thus the concentration of the solution is 0.0731 mole per litre, i.e. $0.0731\,mol\,l^{-1}$.

ITQ 6 The equilibrium system is again

$$Mg(OH)_2(s) \rightleftharpoons Mg^{2+}(aq) + 2OH^-(aq) \qquad (21)*$$

If solid $MgCl_2$ is dissolved in the solution, additional $Mg^{2+}(aq)$ is produced:

$$MgCl_2(s) = Mg^{2+}(aq) + 2Cl^-(aq)$$

so the concentration of $Mg^{2+}(aq)$ is increased. This is an external constraint, which disturbs the equilibrium of Equation 21. According to Le Chatelier's principle, the system reacts so as to lessen the effect of the constraint, that is, so as to reduce the increased concentration of $Mg^{2+}(aq)$. It can do this if the right to left reaction becomes predominant, resulting in a net transfer of $Mg^{2+}(aq)$ and $OH^-(aq)$ into $Mg(OH)_2(s)$. This will lower the concentration of $OH^-(aq)$ below its original value. The lowering goes on until a new position of equilibrium is reached.

ITQ 7 The equilibrium system is

$$HAc(aq) \rightleftharpoons H^+(aq) + Ac^-(aq) \qquad (25)*$$

(i) When NaOH(s) is dissolved in the solution, $OH^-(aq)$ is produced, and then reacts with $H^+(aq)$:

$$H^+(aq) + OH^-(aq) = H_2O(l)$$

This lowering of the hydrogen ion concentration is an external constraint on Equilibrium 25. The system responds by trying to lessen the effect of the constraint, that is, to raise the reduced concentration of $H^+(aq)$. The left to right reaction becomes predominant, and HAc(aq) is transformed into $H^+(aq)$ and $Ac^-(aq)$. The characteristic smell of HAc is therefore diminished. Note that your solution of NaAc(s), which contains much $Na^+(aq)$ and $Ac^-(aq)$, is odourless: $Ac^-(aq)$ does not smell like HAc.

(ii) Sodium acetate is a salt, and is therefore a strong electrolyte:

$$NaAc(s) = Na^+(aq) + Ac^-(aq)$$

When dissolved in acetic acid, it increases the concentration of $Ac^-(aq)$. The system of Equation 25 responds by trying to lower this increased concentration. The right to left reaction becomes dominant and $H^+(aq)$ combines with $Ac^-(aq)$ to form HAc(aq). The hydrogen ion concentration therefore drops.

ITQ 8 As noted in ITQ 7, the addition of sodium hydroxide results in the breakdown of undissociated HAc molecules and a consequent decrease in the smell of the solution.

The $OH^-(aq)$ formed by dissociation of NaOH combines with $H^+(aq)$:

$$H^+(aq) + OH^-(aq) = H_2O(l)$$

This reduces the concentration of $H^+(aq)$, which we write as $[H^+(aq)]$, and therefore lowers the value of

$$\frac{[H^+(aq)][Ac^-(aq)]}{[HAc(aq)]}$$

below the *equilibrium* value of $1.8 \times 10^{-5}\,mol\,l^{-1}$(at 25 °C). In order to return to equilibrium, the system must re-establish this value. The left to right process in Equilibrium 25 becomes dominant:

$$HAc(aq) = H^+(aq) + Ac^-(aq)$$

and some HAc molecules dissociate. The process continues until the consequent reduction in [HAc(aq)] and increase in $[Ac^-(aq)]$ and $[H^+(aq)]$ have increased the lowered value of the fraction back to $1.8 \times 10^{-5}\,mol\,l^{-1}$.

ITQ 9 Statements (a), (d) and (e) are correct; (b) and (c) are incorrect. The addition of $Ca(OH)_2$ raises $[OH^-(aq)]$, but the product $[H^+(aq)][OH^-(aq)]$ must remain at the equilibrium value, so $[H^+(aq)]$ will decrease by the formation of more undissociated water:

$$H^+(aq) + OH^-(aq) = H_2O(l)$$

The increase in the number of water molecules, however, is small compared with the number that are already present. K_w is a constant at constant temperature and so it does not change.

ITQ 10 (a) $10 = 10^1$ so $\log 10 = 1$; $100 = 10^2$ so $\log 100 = 2$; $0.001 = 10^{-3}$ so $\log 0.001 = -3$.

(b) Using your calculator, $\log 4 = 0.6021$; $\log 40 = 1.6021$; $\log 0.4 = -0.3979$.

Notice that

$$\begin{aligned}\log 40 &= \log (10 \times 4)\\ &= \log 10 + \log 4 \text{ (from Equation 38)}\\ &= 1 + 0.6021\\ &= 1.6021\end{aligned}$$

and similarly,

$$\begin{aligned}\log 0.4 &= \log (4 \times 10^{-1})\\ &= \log 4 + \log 10^{-1}\\ &= 0.6021 - 1\\ &= -0.3979\end{aligned}$$

(c) The value is -7.4. If $[H^+(aq)] = 4 \times 10^{-8}\,\text{mol l}^{-1}$

$$\begin{aligned}\log \left\{\frac{[H^+(aq)]}{\text{mol l}^{-1}}\right\} &= \log (4 \times 10^{-8})\\ &= \log 4 + \log 10^{-8}\\ &= 0.6021 - 8\\ &= -7.3979\\ &= -7.4 \text{ (to two significant figures)}\end{aligned}$$

ITQ 11 (i) 10^0, 10^{-1}, 10^{-2} and 10^{-6}, respectively; (ii) 0, -1, -2 and -6, respectively; (iii) 0, 1 and 6, respectively.

The answers to part (ii) are just the powers of 10 in part (i). Since $\text{pH} = -\log ([H^+(aq)]/\text{mol l}^{-1})$, the answers in part (iii) are just minus those in part (ii).

ITQ 12 Orange juice is more acidic than tomato juice; gastric juice is more acidic than vinegar; blood is more acidic than seawater. From the definition of pH, the *lower* the pH, the more acidic the solution.

ITQ 13 Applying Equation 50:

(a) $K = \dfrac{[NO(g)]^2}{[N_2(g)][O_2(g)]}$

(b) $K = \dfrac{[Sn^{4+}(aq)][Fe^{2+}(aq)]^2}{[Sn^{2+}(aq)][Fe^{3+}(aq)]^2}$

(c) $K = \dfrac{[H_2O(g)]^2}{[H_2(g)]^2[O_2(g)]}$

(d) $K = \dfrac{[H_2O(g)]}{[H_2(g)][O_2(g)]^{1/2}}$

Note that for (c) and (d) the two values of K are different, even though they refer to the same reaction—the reaction of hydrogen with oxygen to give steam. This is because in (c) and (d), different balanced equations have been chosen to represent the same reaction, so different equilibrium constants result. This emphasizes the point immediately prior to ITQ 13.

In fact, the equilibrium constant in (c) is the square of that in (d), the experimental values at 25 °C being $33 \times 10^{80}\,\text{mol}^{-1}\,\text{l}$ and $5.74 \times 10^{40}\,\text{mol}^{-1/2}\,\text{l}^{1/2}$. The two expressions are completely equivalent and give exactly the same results for equilibrium concentrations.

SAQ ANSWERS AND COMMENTS

SAQ 1 (i) a; (ii) c, d; (iii) b; (iv) b; (v) a. You are referred to the text.

SAQ 2 (i) Because the formula of calcium hydroxide, $Ca(OH)_2$, yields two hydroxide ions, and that of nitric acid, HNO_3, yields one hydrogen ion, each $Ca(OH)_2$ must react with two HNO_3. In an ionic form,

$$\begin{aligned}Ca^{2+}(aq) + 2OH^-(aq) + 2H^+(aq) + 2NO_3^-(aq)\\ = Ca^{2+}(aq) + 2NO_3^-(aq) + 2H_2O(l)\end{aligned}$$

Eliminating the ions $Ca^{2+}(aq)$ and $NO_3^-(aq)$, which are common to both sides, we get

$$2H^+(aq) + 2OH^-(aq) = 2H_2O(l)$$

This reduces to Equation 11 when divided through by 2.

(ii) By similar reasoning, two LiOH will be required to react with each H_2SO_4. In an ionic form,

$$\begin{aligned}2Li^+(aq) + 2OH^-(aq) + 2H^+(aq) + SO_4^{2-}(aq)\\ = 2Li^+(aq) + SO_4^{2-}(aq) + 2H_2O(l)\end{aligned}$$

On elimination of $Li^+(aq)$ and $SO_4^{2-}(aq)$ and division by 2, Equation 11 again remains.

SAQ 3 (i) Calcium nitrate, $Ca(NO_3)_2$; (ii) lithium sulphate, Li_2SO_4. Combine the cation of the base with the anion of the acid in a ratio that produces electrical neutrality.

SAQ 4 If MgO acts as a base, it must react with and dissolve in water to give some $OH^-(aq)$. With the familiar $Mg^{2+}(aq)$ cation as the other likely product, we get

$$MgO(s) + H_2O(l) = Mg^{2+}(aq) + 2OH^-(aq)$$

Likewise, SO_3 must react with water to yield some

$H^+(aq)$. With the familiar sulphate anion as the other likely product, we get

$$SO_3(s) + H_2O(l) = 2H^+(aq) + SO_4^{2-}(aq)$$

SAQ 5 (i) $0.1\,mol\,l^{-1}$; (ii) $0.2\,mol\,l^{-1}$; (iii) $0.1\,mol\,l^{-1}$.

The molar mass of Na_2SO_4 is

$$[(2 \times 23) + (1 \times 32) + (4 \times 16)]\,g = 142\,g$$

The concentration of the solution is $14.2\,g\,l^{-1}$, which is equal to

$$\frac{14.2}{142}\,mol\,l^{-1} = 0.1\,mol\,l^{-1}$$

The equation for the dissociation is

$$Na_2SO_4(s) = 2Na^+(aq) + SO_4^{2-}(aq)$$

On dissolution therefore, the 0.1 mole of Na_2SO_4 in each litre becomes 0.2 mole of $Na^+(aq)$ and 0.1 mole of $SO_4^{2-}(aq)$.

SAQ 6 (i) A white precipitate of barium sulphate, $BaSO_4$, appears. (ii) No visible reaction: the mixture is a clear colourless solution. (iii) A yellow precipitate of lead iodide, PbI_2, appears. (iv) A white precipitate of lead carbonate, $PbCO_3$, appears. The equations are

(i) $Ba^{2+}(aq) + SO_4^{2-}(aq) = BaSO_4(s)$

(iii) $Pb^{2+}(aq) + 2I^-(aq) = PbI_2(s)$

(iv) $Pb^{2+}(aq) + CO_3^{2-}(aq) = PbCO_3(s)$

See Section 4.3. Answers are obtained by looking for combinations of cations and anions that might give very sparingly soluble compounds. In (ii), all such combinations of the ions Ba^{2+}, NO_3^-, Na^+ and Cl^- ($Ba(NO_3)_2$, $BaCl_2$, $NaCl$, $NaNO_3$) are soluble so there is no visible reaction. In the other cases, note that the ion combination that remains in solution is that of a soluble compound: (i) NaCl, (iii) KNO_3, (iv) $NaNO_3$.

SAQ 7 The concentration of $Ac^-(aq)$ falls and that of HAc(aq) rises. The equilibrium system is

$$HAc(aq) \rightleftharpoons H^+(aq) + Ac^-(aq)$$

HCl is a strong electrolyte and, when it dissolves, it raises the concentration of $H^+(aq)$. The system therefore reacts to lower this increased concentration: $Ac^-(aq)$ reacts with $H^+(aq)$ to form HAc(aq).

SAQ 8 The balanced equation for the reaction is as follows:

$$3H_2(g) + N_2(g) \rightleftharpoons 2NH_3(g)$$

If an equilibrium mixture of H_2, N_2 and NH_3 is washed with a spray of water, much more NH_3 will dissolve than H_2 and N_2. Removing NH_3 in this way constitutes an external constraint on the gaseous equilibrium. According to Le Chatelier's principle, the balance can be restored by further reaction of N_2 and H_2 to form NH_3: in qualitative terms, the equilibrium will be shifted to the right and more ammonia will be formed.

SAQ 9 (i) d; (ii) a; (iii) c; (iv) b; (v) d.

See Section 4.5. From Section 4.2, a saturated solution of $Mg(OH)_2$ is very dilute, but as noted in Section 4.5, all dissolved $Mg(OH)_2$ forms ions. In example (iv), sugar would be better described as a non-electrolyte: unlike acetic acid, any dissociation is negligible. In example (v), potassium iodide is freely soluble in water (Table 2), so a saturated solution will be concentrated. It is also a salt, and all salts are strong electrolytes.

SAQ 10 (i) $[H^+(aq)] = [IO_3^-(aq)] = 0.41\,mol\,l^{-1}$; (ii) 29%.

The acid dissociates thus:

$$HIO_3(aq) = H^+(aq) + IO_3^-(aq)$$

and, from Table 3, the acid dissociation constant, K_a, is,

$$K_a = \frac{[H^+(aq)][IO_3^-(aq)]}{[HIO_3(aq)]} = 0.17\,mol\,l^{-1}$$

The acid is added to water until $[HIO_3(aq)] = 1\,mol\,l^{-1}$. Those molecules that have dissociated have yielded equal numbers of the ions $H^+(aq)$ and $IO_3^-(aq)$, so

$$[H^+(aq)] = [IO_3^-(aq)]$$

Therefore,

$$\frac{[IO_3^-(aq)]^2}{1\,mol\,l^{-1}} = 0.17\,mol\,l^{-1}$$

$$[IO_3^-(aq)]^2 = 0.17\,mol^2\,l^{-2}$$

$$[IO_3^-(aq)] = 0.41\,mol\,l^{-1}$$

Thus, of the total acid added to each litre of solution, 1 mole remains undissociated, and 0.41 mole has dissociated. That is, 1.41 moles were added to each litre, of which 0.41 mole, or 29%, is dissociated into ions.

Note that the percentage of dissociation is much greater than in the analogous example of acetic acid in Section 5.1. This is because K_a is much greater.

SAQ 11
$[HX(aq)] = [H^+(aq)] = [X^-(aq)] = 0.05\,mol\,l^{-1}$; $K_a = 0.05\,mol\,l^{-1}$.

The dissociation equilibrium is

$$HX(aq) \rightleftharpoons H^+(aq) + X^-(aq)$$

If the total concentration of dissolved acid is $0.1\,mol\,l^{-1}$, and half of the acid molecules remain undissociated in the solution, then $[HX(aq)] = 0.05\,mol\,l^{-1}$. The other half of the acid molecules break down into equal numbers of $H^+(aq)$ and $X^-(aq)$ ions, so $[H^+(aq)] = [X^-(aq)] = 0.05\,mol\,l^{-1}$. The acid dissociation constant is then obtained as:

$$K_a = \frac{[H^+(aq)][X^-(aq)]}{[HX(aq)]}$$

$$= \frac{(0.05\,mol\,l^{-1}) \times (0.05\,mol\,l^{-1})}{0.05\,mol\,l^{-1}}$$

$$= 0.05\,mol\,l^{-1}$$

From Table 3, K_a for $HNO_2(aq)$ is 4.5×10^{-4} mol l^{-1} or 0.000 45 mol l^{-1}, that is, much smaller than 0.05 mol l^{-1}. This suggests that in a solution of HNO_2 of concentration 0.1 mol l^{-1}, a lot more than half the dissolved acid must be in an undissociated state.

SAQ 12 Only (a) is correct.

In aqueous solution, HNO_3 is completely dissociated into ions. As this is not true of solutions of HNO_2 (SAQ 11), this must mean that HNO_3 is the stronger acid, so (b) is incorrect. This is true whatever concentrations of the two acids are being compared, because acid strength is related to the *proportion* of total acid that dissociates, not to the total concentration, or even to the concentration of $H^+(aq)$ in solution. Some solutions of HNO_2 will have larger values of $[H^+(aq)]$ than some containing very much lower concentrations of HNO_3, but statement (b) is always incorrect.

In this case the HNO_2 is twice as concentrated as the HNO_3 but, as you found in SAQ 11, well over half the HNO_2 is undissociated. This means that $[H^+(aq)]$ must be greater in the HNO_3 solution, so (c) is incorrect.

SAQ 13 A more accurate figure is one in 560 million. In Section 6.1, the concentration of water molecules in pure water *assuming no ionization* is calculated to be 55.6 mol l^{-1}, and $[H^+(aq)] = 1 \times 10^{-7}$ mol l^{-1}. Every water molecule that ionizes produces one $H^+(aq)$, so

$$[\text{ionized } H_2O] = [H^+(aq)] = 1.0 \times 10^{-7}\,\text{mol l}^{-1}$$

The concentration of ionized water molecules is thus negligible when compared with 55.6 mol l^{-1}, so the latter figure will serve for the concentration of water molecules left un-ionized. Thus in one litre there are 55.6 moles of un-ionized and 1.0×10^{-7} mole of ionized water molecules: the ratio is 560 million (to two significant figures).

SAQ 14 $[H^+(aq)] = 1.0 \times 10^{-4}$ mol l^{-1}; pH = 4; $[OH^-(aq)] = 1.0 \times 10^{-10}$ mol l^{-1}.

Nitric acid is strong, and every mole dissociates to give one mole of $[H^+(aq)]$. Assuming that the final volume of the solution is 10 litres, the dissociation of the HNO_3 yields

$$[H^+(aq)] = \frac{0.001\,\text{mol}}{10\,\text{l}} = 1.0 \times 10^{-4}\,\text{mol l}^{-1}$$

This is far greater than $[H^+(aq)]$ in pure water (10^{-7} mol l^{-1}), so we can ignore any contribution from the solvent and take it to be the real value of $[H^+(aq)]$. Then pH $= -\log[10^{-4}] = -(-4) = 4$. Using the ion product of water,

$$[OH^-(aq)] = \frac{1.0 \times 10^{-14}\,\text{mol}^2\,\text{l}^{-2}}{[H^+(aq)]}$$

$$= \frac{1.0 \times 10^{-14}\,\text{mol}^2\,\text{l}^{-2}}{1 \times 10^{-4}\,\text{mol l}^{-1}}$$

$$= 1.0 \times 10^{-10}\,\text{mol l}^{-1}$$

SAQ 15 (i) $[OH^-(aq)] = 0.040$ mol l^{-1}; (ii) $[H^+(aq)] = 2.5 \times 10^{-13}$ mol l^{-1}; (iii) pH = 12.6.

First, recall that the solubility of a solid represents the concentration of the saturated solution. The dissociation of $Ca(OH)_2$ in aqueous solution can be represented as

$$Ca(OH)_2(s) = Ca^{2+}(aq) + 2OH^-(aq)$$

Assuming that all dissolved material is in the form of $Ca^{2+}(aq)$ and $OH^-(aq)$ ions, that is that $Ca(OH)_2$ is a strong electrolyte, then as each mole of $Ca(OH)_2$ yields 2 moles of $OH^-(aq)$,

$$[OH^-(aq)] = 2 \times 0.020\,\text{mol l}^{-1}$$

$$= 0.040\,\text{mol l}^{-1}$$

To calculate the pH, $[H^+(aq)]$ must be calculated from the ion product of water, at 25 °C:

$$K_w = [H^+(aq)][OH^-(aq)] = 1.0 \times 10^{-14}\,\text{mol}^2\,\text{l}^{-2}$$

$$[H^+(aq)] = \frac{10^{-14}}{0.040}\,\text{mol l}^{-1} = 2.5 \times 10^{-13}\,\text{mol l}^{-1}$$

From the definition of $\text{pH} = -\log\left\{\frac{[H^+(aq)]}{\text{mol l}^{-1}}\right\}$

$$\text{pH} = -\log(2.5 \times 10^{-13})$$

$$= -(\log(2.5) + \log(10^{-13}))$$

Using your calculator,

$$\text{pH} = -(0.3979 - 13) \approx 12.6$$

As you should have anticipated, the solution is strongly basic.

SAQ 16 (i) $Cl_2O_7 + H_2O(l) = 2HClO_4(aq)$

$$HClO_4(aq) = H^+(aq) + ClO_4^-(aq)$$

(ii) HClO (hypochlorous acid) is the weaker of the two acids:

$$HClO(aq) \rightleftharpoons H^+(aq) + ClO^-(aq)$$

You may be interested to know that for HClO, K_a is only 2.9×10^{-8} mol l^{-1} (Table 3).

You were told that one of the acids is strong and the other is weak. But from Section 6.3 we know that the higher oxide is the more acidic. Thus Cl_2O_7 must form the strong acid and Cl_2O the weak one. Note however that both oxides are acidic, as are most oxides of non-metals (Section 6.3).

SAQ 17 (i) $K_a = 1.78 \times 10^{-5}$ mol l^{-1}; (ii) the ratio of un-ionized to ionized acetic acid molecules is 56.2.

(i) In the solution, pH = 3.00, so

$$\log\left\{\frac{[H^+(aq)]}{\text{mol l}^{-1}}\right\} = -3.00$$

$$[H^+(aq)] = 10^{-3}\,\text{mol l}^{-1}$$

As every HAc(aq) yields one $H^+(aq)$ and one $Ac^-(aq)$:

$$[H^+(aq)] = [Ac^-(aq)] = 10^{-3}\,\text{mol l}^{-1}$$

Thus in every litre there is 10^{-3} or 0.001 mole of ionized acid. As the total amount of acid in each litre is 0.0572 mole, each litre contains (0.0572 − 0.001) mole of un-ionized acid. So,

$$[HAc(aq)] = 0.0562\,mol\,l^{-1}$$

Now K_a can be evaluated:

$$K_a = \frac{[H^+(aq)][Ac^-(aq)]}{[HAc(aq)]}$$

$$= \frac{(0.001)^2\,mol^2\,l^{-2}}{0.0562\,mol\,l^{-1}} = 1.78 \times 10^{-5}\,mol\,l^{-1}$$

This is very close to the value in Table 3.

(ii) In each litre there are 0.0562 mole of un-ionized, and 0.001 mole of ionized acid. The ratio is therefore 56.2.

SAQ 18 From Equation 50,

$$K = \frac{[HI(g)]^2}{[H_2(g)][I_2(g)]}$$

$$= \frac{(1.70)^2\,mol^2\,l^{-2}}{(0.35\,mol\,l^{-1})(0.15\,mol\,l^{-1})} \approx 55$$

This is close to the quoted figure.

SAQ 19 According to Equation 50, the equilibrium constant for the reaction

$$N_2(g) + 3H_2(g) \rightleftharpoons 2NH_3(g)$$

is given by

$$K = \frac{[NH_3(g)]^2}{[N_2(g)][H_2(g)]^3}$$

If $NH_3(g)$ is removed from the equilibrium system, $[NH_3(g)]$ is decreased, but at equilibrium, the right-hand side of the above equation cannot fall below the equilibrium value. Consequently, the system responds in the way discussed in SAQ 8: more N_2 and H_2 react to form NH_3 until the equilibrium value of the expression is restored. In this context, the predictions based on Le Chatelier's principle are just a direct consequence of the constancy of K at a given temperature.

SAQ 20 (i) 0.944 mole of $NH_3(g)$; (ii) 0.475 mole of $N_2(g)$ and 1.426 moles of $H_2(g)$.

At equilibrium, $[N_2(g)] = 0.0033\,mol\,l^{-1}$ and $[H_2(g)] = 0.0100\,mol\,l^{-1}$, so

$$\frac{[NH_3(g)]^2}{(3.3 \times 10^{-3}\,mol\,l^{-1})(1 \times 10^{-2}\,mol\,l^{-1})^3} = 2.70 \times 10^8\,mol^{-2}\,l^2$$

$$[NH_3(g)]^2 = 8.91 \times 10^{-1}\,mol^2\,l^{-2}$$

$$[NH_3(g)] = 0.944\,mol\,l^{-1}$$

If we divide the formation equation by 2:

$$\tfrac{1}{2}N_2(g) + \tfrac{3}{2}H_2(g) = NH_3(g)$$

we see that every mole of ammonia is formed from $\frac{1}{2}$ mole of N_2 and $1\frac{1}{2}$ moles of H_2. Thus the 0.944 mole of ammonia in the litre bulb was formed from 0.472 mole of N_2 and 1.416 moles of H_2. If these amounts are added to the 0.0033 mole of N_2 and 0.01 mole of H_2 present at equilibrium, the total amount pumped in must have been 0.475 mole of N_2 and 1.426 moles of H_2.

INDEX FOR UNIT 15

THE OPEN UNIVERSITY
A SCIENCE FOUNDATION COURSE

UNIT 16 CHEMICAL ENERGETICS

THE OPEN UNIVERSITY PRESS

COMBUSTION REACTION

EXOTHERMIC REACTION

STUDY GUIDE

This Unit has just two main components: the text and a television programme, 'Energy and rockets'. You will find notes about this programme in Section 8. You should try to read up to the end of Section 3 before watching although this is not absolutely necessary to understand the programme.

As with Unit 15 there is a CALCHEM program available for this Unit; it provides exercises and examples on Sections 2, 3, and 5.

1 INTRODUCTION

In Unit 15 you were mainly concerned with questions like: what substances are formed in a reaction? How far will the reaction go—what is the equilibrium position? What happens when the equilibrium is disturbed in some way? This Unit examines a slightly different aspect of chemical reactions.

How do you heat your home and cook your food? What fuel do you use: is it natural gas or oil or perhaps a solid fuel, like coal or coke—or do you use electricity? Leaving aside electricity for a moment, what do you do with the fuel? Of course, you *burn* it in the oxygen in the air. More precisely, the characteristic and useful property of all these so-called 'fossil' fuels lies in their reaction with oxygen: chemical energy is *released* when they are burnt in oxygen. For example, coke is mainly carbon, and the combustion of carbon in oxygen, which can be represented by the chemical equation

$$C(s) + O_2(g) = CO_2(g) \qquad (1)$$

results not only in the formation of gaseous carbon dioxide, but also in the release of energy in the form of heat and light. In the home, energy released in **combustion reactions** like this is generally used *directly* to heat something else; the air in your living room, the water in a saucepan on the stove or that circulating through the radiators; the food in your oven. At the time of writing, however, most forms of transport depend *indirectly* on the release of *chemical* energy in similar combustion reactions, as does the generation of most of our electricity.

But combustion in oxygen is just one sort of chemical reaction, and chemical reactions in general are accompanied by energy changes though they may be considerably smaller than in the examples cited above. We begin this Unit by examining the energetics of chemical reactions in this broader context, before returning (in Section 4) to take a closer look at chemical fuels.

FIGURE 1 The Ronan Point disaster (1968), which was sparked off by the explosive reaction between methane and oxygen.

You will also be concerned with a rather different, but related, question. When you turn on a gas tap, the gas does not immediately burst into flames. Indeed, a mixture of natural gas and air in your kitchen, or in any other container, will remain essentially unchanged for as long as you care to wait. It is common knowledge, however, that a tiny spark or match flame is sufficient to initiate a violent chemical reaction (Figure 1). The major ingredient of natural gas is methane, which has the formula CH_4. It burns in oxygen to give carbon dioxide and water:

$$CH_4(g) + 2O_2(g) = CO_2(g) + 2H_2O(g) \qquad (2)$$

In the absence of a spark or flame, this reaction occurs so *slowly* that we do not even notice it.

This, then, is the question we shall examine in Section 5 of the Unit—the speed or *rate* of a chemical reaction, and how this is influenced by the reaction conditions. In many ways, this study is a necessary complement to the ideas developed in Unit 15. You have seen that a reaction can go only as far as the position of equilibrium dictates: later in this Unit you will examine some of the factors that govern *how fast* this equilibrium position is attained. To draw together these complementary strands, the Unit closes (Section 7) by examining an industrial process that has had a profound impact on our society—the Haber process for 'fixing' molecular nitrogen from the atmosphere.

2 EXOTHERMIC AND ENDOTHERMIC CHEMICAL REACTIONS

All of the combustion reactions mentioned in the previous Section release energy in the form of heat: such reactions are described as being **exothermic** (from the Greek *exo* meaning outside and *therme* meaning heat). One of the examples cited earlier is the combustion of coke in oxygen, which can be simplified as:

$$C(s) + O_2(g) = CO_2(g) \qquad (1)^*$$

This reaction obviously results in the release of a considerable amount of energy. By comparison, you would probably agree that dissolving solid sodium hydroxide in water, Equation 3, is a rather unexciting process:

$$NaOH(s) = Na^+(aq) + OH^-(aq) \qquad (3)$$

However, as you may have noticed during Experiment 2 of Unit 15, when NaOH dissolves in water, the tube feels hotter. In other words, the process in Equation 3 results in a (fairly modest) temperature rise. This observation indicates that the dissolving process for NaOH is also exothermic; that is, it is accompanied by the release of energy, which heats up the solution.

The actual temperature rise produced by an exothermic reaction depends on many factors, for instance the conditions under which it takes place and how fast it goes. For example, the reaction between iron and oxygen to form iron oxide is an exothermic process. Nevertheless, your car does not get noticeably hotter as it slowly rusts by reaction with the oxygen in the atmosphere! More generally, if an exothermic reaction is carried out in an insulated container (such as a Thermos flask), the energy liberated will be slow to escape and the temperature may then become quite high. If the *same* reaction is performed under conditions in which heat loss is encouraged, a lower temperature will be produced. In addition, fast exothermic reactions tend to produce high temperatures because there is insufficient time for the energy released to be dissipated. (You will see some spectacular examples in the TV programme 'Energy and rockets'.) Slow exothermic reactions generally produce smaller temperature rises than fast ones.

Now, throughout the discussion of chemical equilibrium in Unit 15, you were interested in reactions taking place at *constant temperature*: you know how to write down the equilibrium constant for a reaction under these conditions. Moreover, you have seen that the equilibrium position, as characterized by the size of the equilibrium constant, depends on the temperature. For these reasons, chemists are mainly interested in the energetics of reactions at constant temperature. The one universal statement that it is possible to make about exothermic reactions is that if the products of such a reaction are finally obtained *at the same temperature* as the reactants, then heat must have left the reaction vessel. This is illustrated schematically in Figure 2.

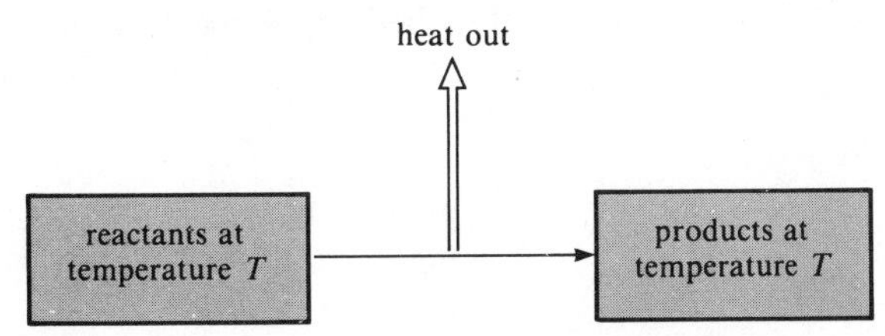

FIGURE 2 Schematic representation of an exothermic reaction at constant temperature.

□ When potassium nitrate, KNO_3, dissolves in water, the tube gets colder. This process can be represented by the following equation:

$$KNO_3(s) = K^+(aq) + NO_3^-(aq) \qquad (4)$$

According to the discussion above, is this an exothermic process?

ENDOTHERMIC REACTION

ENTHALPY OF REACTION ΔH

PHASE CHANGE

■ No. The process represented by Equation 4 results in a *fall* in the temperature. According to the discussion above, heat must be *added* (not lost) in order to keep the temperature constant, so the process cannot be exothermic.

Reactions like this are described as being **endothermic** (from the Greek *endo* meaning within). Figure 3 shows a schematic picture of an endothermic process at constant temperature.

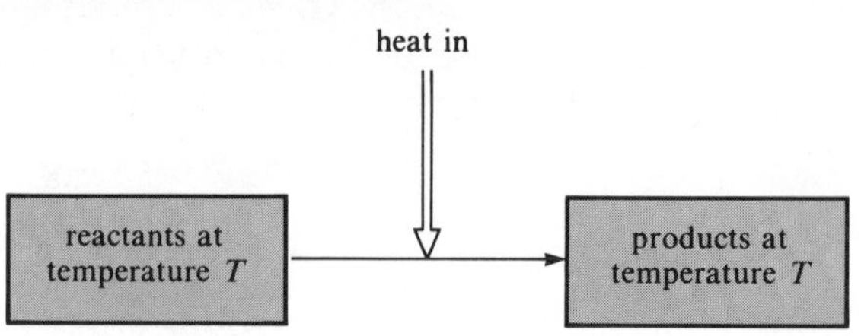

FIGURE 3 Schematic representation of an endothermic reaction at constant temperature.

Now, heat transfer is generally symbolized by the letter Q. *By convention*, the value of Q for any process is defined as positive (greater than zero, >0) if heat is added to the system of interest (the reaction mixture, in this case), but negative (less than zero, <0) if heat is released to the surroundings.

□ According to this convention, what is the sign of Q for an exothermic reaction? What about an endothermic process?

■ To remain at constant temperature, an exothermic reaction must release heat to the surroundings, and vice versa for an endothermic reaction. The sign convention for heat transfer Q, suggests that:

for an exothermic reaction: Q is negative, $Q < 0$

for an endothermic reaction: Q is positive, $Q > 0$ (5)

As well as being at constant temperature, most reactions of interest to chemists, and indeed many natural processes as well, take place at constant pressure. Reactions you carry out in a beaker open to the atmosphere are essentially at constant pressure. Because of this, chemists have chosen to define a property that expresses the heat transferred under this condition. Thus, if a reaction takes place *at constant temperature and pressure*, the heat transferred is called the **enthalpy of reaction**, and is denoted by the symbol ΔH. Under these conditions,

$$\Delta H = Q \tag{6}$$

The symbol Δ (Greek capital delta) means a change in some physical quantity, in this case a change in *enthalpy*; that is

$$\Delta H = H(\text{products}) - H(\text{reactants}) \tag{7}$$

As implied above, enthalpy is just the most convenient kind of 'energy quantity' to describe changes at constant pressure, and throughout this Unit the terms enthalpy and energy will frequently be used interchangeably. Thus, you can read Equations 6 and 7 as saying that the heat released (or absorbed) by a reaction is a measure of the difference in energy between reactants and products under the same conditions of temperature and pressure.

□ According to the definitions in expressions 5 and 6, what will be the *signs* of ΔH for an exothermic reaction and for an endothermic reaction?

■ For an exothermic reaction: ΔH is negative, $\Delta H < 0$

For an endothermic reaction: ΔH is positive, $\Delta H > 0$ (8)

In summary, then, a reaction is exothermic if the enthalpy change ΔH at constant temperature is negative: according to Equation 7, for ΔH to be negative the products must be of *lower* energy than the reactants. By the law of conservation of energy (Unit 9), the energy lost must go somewhere, and you have seen that it results in heat being evolved. For an endothermic reaction, the situation is exactly the reverse. So the relative energies of reactants and products in exothermic and endothermic reactions can be represented as shown schematically in Figure 4.

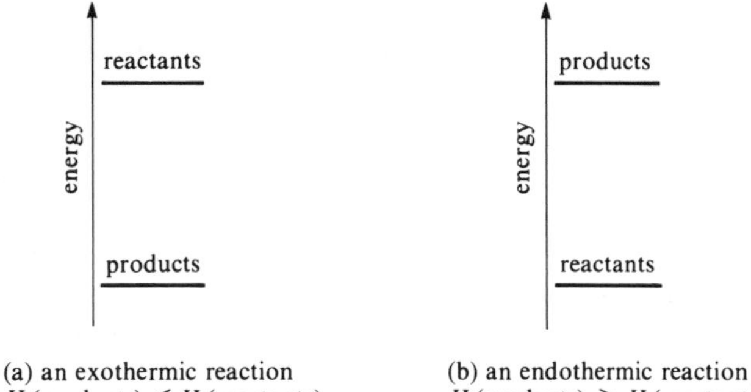

(a) an exothermic reaction
H (products) $<$ H (reactants)
so $\Delta H < 0$

(b) an endothermic reaction
H (products) $>$ H (reactants)
so $\Delta H > 0$

FIGURE 4 Schematic representation of (a) an exothermic reaction and (b) an endothermic reaction, defined in terms of enthalpy changes at constant temperature and pressure.

To make these ideas more concrete, consider a couple of familiar examples. You probably know that pure water boils at 100 °C at normal atmospheric pressure. But does it continue to boil at this temperature as the heating is continued, or does the temperature of the water rise? If you have a suitable thermometer, you may like to try this experiment. You should find that the temperature remains constant until the last drop of water disappears. In other words, heat is supplied, but the temperature does not change.

What does happen?

The liquid (denoted by 'l') water disappears as gaseous water (steam) is formed. It seems that the *constant-temperature* process represented by Equation 9 *absorbs* energy: it is an endothermic process.

$$H_2O(l) = H_2O(g) \tag{9}$$

☐ What will be the *sign* of the enthalpy change for the process in Equation 9?

■ According to the discussion above, ΔH should be positive.

It is: experimentally, it is found to have the following value at 100 °C:

$$\Delta H = +43\,300\,\text{J mol}^{-1}$$

$$= +43.3\,\text{kJ mol}^{-1}$$

This example can be used to make two important generalizations. First, notice the unit, energy *per mole*: it requires 43.3 kJ to vaporize one mole (18 g) of water, 86.6 kJ for two moles, etc. In general, *the enthalpy change for a reaction depends on the amount(s) of reactant(s) consumed.*

ITQ 1 Suppose that you have a litre of pure water, which has just reached boiling temperature. How much energy is required to vaporize half this volume? Assume that 1 cm³ of water has a mass of 1 g.

$43 \cdot 3 \times 10^3$ FOR 18g
2405·56 FOR 1g
× 50 for 500cm³
~~120277·78 J~~
1202·78 kJ

Secondly, the process represented by Equation 9 is one example of a class of similar processes known collectively as **phase changes**, sometimes called phase transitions. The term 'phase' simply means the physical states of a substance, be it solid, liquid or gas. In general, the transitions from solid to liquid (ice melting, for example), from liquid to gas (as you saw above) and, rather less common, from solid to gas,* all require energy: they are all endothermic processes.

* An example is provided by dry ice, which is solid carbon dioxide: this *sublimes* in air, that is it forms gaseous CO_2 directly. The fact that this process is endothermic accounts, in part, for the use of solid CO_2 as a coolant. It absorbs heat as it sublimes, and hence cools its surroundings. (It's also cold to start with, of course.) A second example is the sublimation of solid iodine, which probably caused iodine to leak from the bottle in your Experiment Kit.

Endothermic: the heat is absorbed from the skin.

ITQ 2 If you dab ether on your skin, it feels cold. Does this effect accord with the generalization made above?

The fact that these transitions are endothermic is consistent with the ideas about chemical bonding in Sections 6.4, 6.6 and 7 of Units 13–14. For example, it evidently requires energy to overcome the forces that hold the ions or molecules in a solid in their fixed positions, and hence to allow the relative freedom of movement that characterizes liquids. Similarly, it requires energy to overcome the forces that hold together the molecules in a liquid—in a drop of water, for instance—and hence to allow the complete freedom of movement that characterizes a gas.

Consider now another exothermic process, the reaction between hydrogen and chlorine gases, which you met in Unit 15:

$$H_2(g) + Cl_2(g) = 2HCl(g) \qquad (10)$$

It is found experimentally that the reaction of one mole (2 g) of H_2 with one mole (71 g) of Cl_2, at constant temperature and pressure, releases 184 kJ of heat.

□ What is the enthalpy change for Equation 10?

■ $\Delta H = -184\,\text{kJ mol}^{-1}$

Notice that the enthalpy change is for the reaction *as written in Equation 10*: the value of ΔH represents the heat transferred when *one* mole of H_2 reacts with *one* mole of Cl_2 to give *two* moles of HCl. The 'mol^{-1}' in the value, $-184\,\text{kJ mol}^{-1}$, here implies that 184 kJ of heat are released *per mole of Equation 10: it refers to the molar quantities implied by a particular equation.* Thus if we choose to represent the reaction by the equation

$$\tfrac{1}{2}H_2(g) + \tfrac{1}{2}Cl_2(g) = HCl(g) \qquad (11)$$

the molar quantities implied by Equation 11 are halved relative to Equation 10.

□ What is the enthalpy change for Equation 11?

■ $\Delta H = -92\,\text{kJ mol}^{-1}$; half the figure for Equation 10. The value tells us that 92 kJ of heat are released when *half* a mole of H_2 reacts with *half* a mole of Cl_2 to give *one* mole of HCl.

You should always remember that a value of ΔH refers to a particular chemical equation. The values that we have been quoting can be determined experimentally, although we do not go into that here. However, you will see in the next Section how enthalpy changes for certain types of reaction can be estimated in a rather simple way.

SUMMARY OF SECTION 2

1 The enthalpy change ΔH for a chemical reaction is defined as the heat transferred under conditions of constant temperature and pressure.

2 Reactions can be classified according to the sign of ΔH, as either exothermic (ΔH negative) or endothermic (ΔH positive). At constant temperature, an exothermic reaction releases heat; the products are of lower energy than the reactants. For an endothermic reaction, the situation is exactly the reverse.

3 The enthalpy change for a reaction depends on the amount(s) of reactant(s) consumed.

4 In the case of phase changes, one can predict whether the process will be exothermic or endothermic.

SAQ 1 (a) Given the enthalpy change for the following reaction:

$$H_2(g) + Br_2(l) = 2HBr(g); \quad \Delta H = -72\,\text{kJ mol}^{-1}$$

which of the following statements are true?

(i) The reaction is endothermic.

(ii) The reaction is exothermic.

(iii) If the reaction is performed at constant temperature, heat will be released.

(iv) If the reaction is performed at constant temperature, heat will be absorbed.

(b) If 1 g of H_2 reacts completely with 80 g of Br_2, and the temperature is held constant, how much heat will be transferred? (The relative atomic masses are H = 1 and Br = 80.)

SAQ 2 According to the discussion in Section 2, which of the processes (a) to (d) should be exothermic and which endothermic? Are there any of the processes for which it is not possible to decide on the basis of the discussion in Section 2?

(a) $I_2(s) = I_2(g)$

(b) $NaCl(s) = Na^+(aq) + Cl^-(aq)$

(c) $CH_4(g) + 2O_2(g) = CO_2(g) + 2H_2O(g)$

(d) $H_2O(l) = H_2O(s)$

SAQ 3 A scald from steam is generally more serious than one from contact with an equivalent amount of hot, even boiling, water. Can you suggest why this should be so?

3 ENTHALPY CHANGES ON THE MOLECULAR LEVEL

The previous Section introduced the vocabulary that chemists use when describing the energy changes associated with reactions. But why is it that some reactions are exothermic and others endothermic? Where does the energy come from or go to? In this Section we shall attempt to interpret the overall energy change for a reaction in terms of events on the molecular level.

Consider the simple gas reaction mentioned in Section 2:

$$H_2(g) + Cl_2(g) = 2HCl(g) \qquad (10)^*$$

As you saw in Units 13–14, gaseous hydrogen, chlorine and hydrogen chloride are all *covalent* substances: in the gas phase, each is composed of discrete diatomic molecules.

□ Write down Lewis structures for the molecules H_2, Cl_2 and HCl.

■ H:H ⁝Cl⁝Cl⁝ ⁝Cl⁝H (Lewis structures with electrons shown as × and •)

In each case, the molecule is held together by a single covalent bond, formed by sharing a pair of electrons, which, as you will remember from Units 13–14, can be represented more simply by a dash, as in H—H, for example.

□ Look again at Equation 10. What changes take place on the molecular level when the reaction occurs?

■ On the molecular level, the net effect of this reaction is to break H—H and Cl—Cl bonds, and form H—Cl bonds.

BOND DISSOCIATION ENERGY

To get more practice with writing Lewis structures for simple gaseous molecules, try the following ITQ.

ITQ 3 List the bonds broken and formed in the following reaction:

$$3H_2(g) + N_2(g) = 2NH_3(g)$$

Start by writing the Lewis structures for an N_2 and an NH_3 molecule. (The hydrogen and nitrogen atoms have electronic configurations $1s^1$ and $1s^2 2s^2 2p^3$, respectively.)

H H

In general, chemical reactions of gaseous covalent compounds like those mentioned above involve breaking up reactant molecules and forming product molecules. The theme developed in this Section is that the energy released or absorbed during such a reaction results from a reordering of the way atoms are bound together, that is, from the breaking and making of chemical bonds.

3.1 BOND-BREAKING: BOND ENERGIES

Consider a molecule of hydrogen, H_2. Now, under normal conditions, for example in hydrogen gas at room temperature and atmospheric pressure, molecules of H_2 do not simply 'fall apart' into hydrogen atoms. The pairs of hydrogen atoms are held together by chemical bonds, the result of powerful forces between the atomic pairs. It is, therefore, not surprising that it requires an *input* of energy to tear the two hydrogen atoms apart, that is, to 'break' the H—H bond. The energy required is usually called the **bond dissociation energy** and denoted by the symbol D. Here we shall call it by the simpler name, *bond energy*. The particular bond under consideration is included in brackets after the D. So, in this case

$$D(\text{H—H}) = 7.24 \times 10^{-19}\,\text{J per molecule}$$

However, the energy required to break one mole of H—H bonds is a more useful quantity. This can be obtained by multiplying 7.24×10^{-19} J per molecule by the number of hydrogen molecules in one mole. This number is the Avogadro constant ($6.022 \times 10^{23}\,\text{mol}^{-1}$); that is

$$D(\text{H—H}) = 7.24 \times 10^{-19}\,\text{J} \times 6.022 \times 10^{23}\,\text{mol}^{-1}$$
$$= 4.36 \times 10^{5}\,\text{J mol}^{-1}, \text{ or } 436\,\text{kJ mol}^{-1}$$

This, then, is the energy required to break up one mole of molecular hydrogen to atoms.*

To generalize this idea a little: the bond energy associated with each chemical bond is unique to that bond. In the case of gaseous diatomic molecules like H_2, this corresponds to the energy required to decompose the molecule into gaseous atoms. A few examples are given in Table 1.

□ Write a Lewis structure for the O_2 molecule.

■ As you saw in Units 13–14, oxygen has 6 electrons in its outermost shell. In O_2, each oxygen atom can attain a noble gas structure by sharing *two* pairs of electrons, as:

$$\overset{\times\ \times}{\underset{\times\ \times}{\times\text{O}\times}}\overset{\times}{\underset{\times}{\times}}\text{O}\times \quad \text{or} \quad \text{O=O}$$

Enter this last structure in the space provided in Table 1.

* With the simple representation H—H, it is tempting to think of the molecule as two atoms held together at a fixed distance. But this is not quite right. Spectroscopic measurements on molecules like H_2 indicate that the two atoms are not held rigidly apart at a fixed distance. This is why you may see the molecule represented as two masses (the H atoms) connected by a spring (Figure 5). The spring is flexible, and allows the two masses to move towards and away from each other. In other words, the molecule vibrates about some average separation. In terms of this 'spring model', you can think of the bond energy as the energy required to stretch the spring from its average length to the point where it snaps.

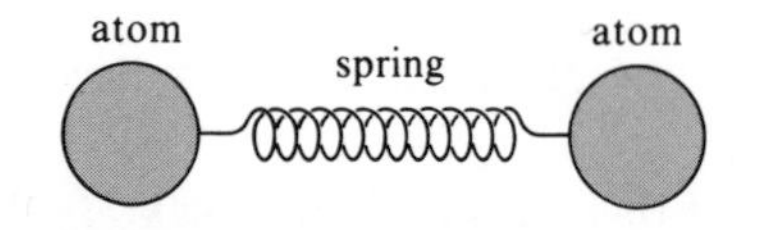

FIGURE 5 Ball-and-spring model of a diatomic molecule.

TABLE 1 Bond energies for selected diatomic molecules at 25 °C

Molecule	Bond	D/kJ mol^{-1}
H_2	H—H	436
F_2	F—F	158
Cl_2	Cl—Cl	244
Br_2	Br—Br	193
I_2	I—I	151
HF	H—F	568
HCl	H—Cl	432
HBr	H—Br	366
HI	H—I	299
O_2	O=O	498
N_2	N≡N	945

Notice that in the cases of O_2 and N_2 (whose Lewis structure you used to answer ITQ 3), dissociation of the diatomic molecule corresponds to breaking *multiple* bonds, a double bond in O_2 and a triple bond in N_2.

The values in Table 1 indicate, as you would expect, that the bond energy is a measure of the 'strength' of a particular bond (like the strength of the spring in Figure 5). For example, the variation of D/kJ mol^{-1} in the series N_2 (945), O_2 (498), F_2 (158) clearly illustrates the greater strength of multiple bonds over single bonds in this group of elements, which are neighbours in the same row of the Periodic Table. The weakness of the F—F bond accounts, at least in part, for the extreme reactivity of fluorine. (The TV programme 'Energy and rockets' shows some spectacular reactions involving fluorine.)

3.2 THE USE OF BOND ENERGIES

We said earlier that reactions of covalent compounds such as those mentioned in Section 3.1 involve breaking bonds and making new ones. The additional implication was that the energy changes associated with such reactions result from changes in the way atoms are bound together. To examine this idea more closely, consider again the simple gas reaction

$$H_2(g) + Cl_2(g) = 2HCl(g) \qquad (10)^*$$

for which the enthalpy change is

$$\Delta H = -184\,\text{kJ mol}^{-1}$$

The reaction is exothermic. This is shown schematically in Figure 6, where the energy of the reactants ($H_2 + Cl_2$) is taken arbitrarily to be zero. This convenient choice of the zero level is permissible because only energy *changes* are significant.

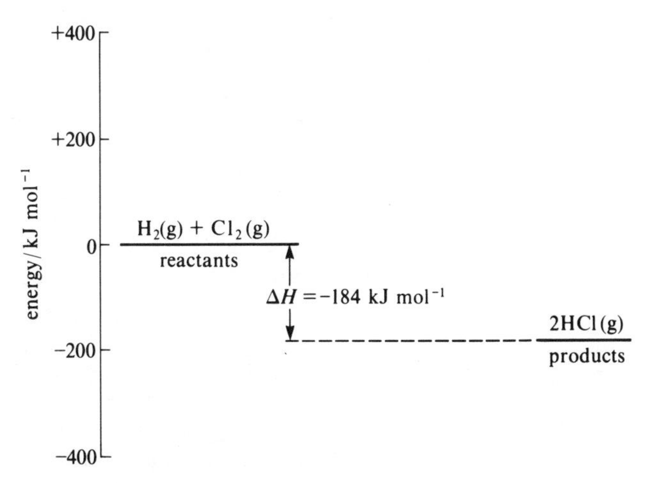

FIGURE 6 Energy diagram for the reaction $H_2(g) + Cl_2(g) = 2HCl(g)$.

HESS'S LAW

Now it is possible to think of this reaction proceeding by the following series of steps. First H_2 molecules and Cl_2 molecules are broken up:

$$H_2(g) = 2H(g) \tag{12}$$

and

$$Cl_2(g) = 2Cl(g) \tag{13}$$

Then the H and Cl atoms so formed are recombined to make HCl:

$$2H(g) + 2Cl(g) = 2HCl(g) \tag{14}$$

□ What is the enthalpy change for the process in Equation 12?

■ According to the discussion in Section 3.1, breaking the H—H bond in H_2, requires an *input* of energy equivalent to the bond energy: it is an endothermic process, so taking the appropriate value from Table 1:

$$\Delta H(12) = D(\text{H—H}) = +436\,\text{kJ mol}^{-1}$$

Similarly, from Table 1 the enthalpy change for Reaction 13 is

$$\Delta H(13) = D(\text{Cl—Cl}) = +244\,\text{kJ mol}^{-1}$$

These steps, together with the overall reaction, are shown schematically in Figure 7. But what about the process in Equation 14, indicated by a question mark on the right-hand side of Figure 7? What kind of process is this?

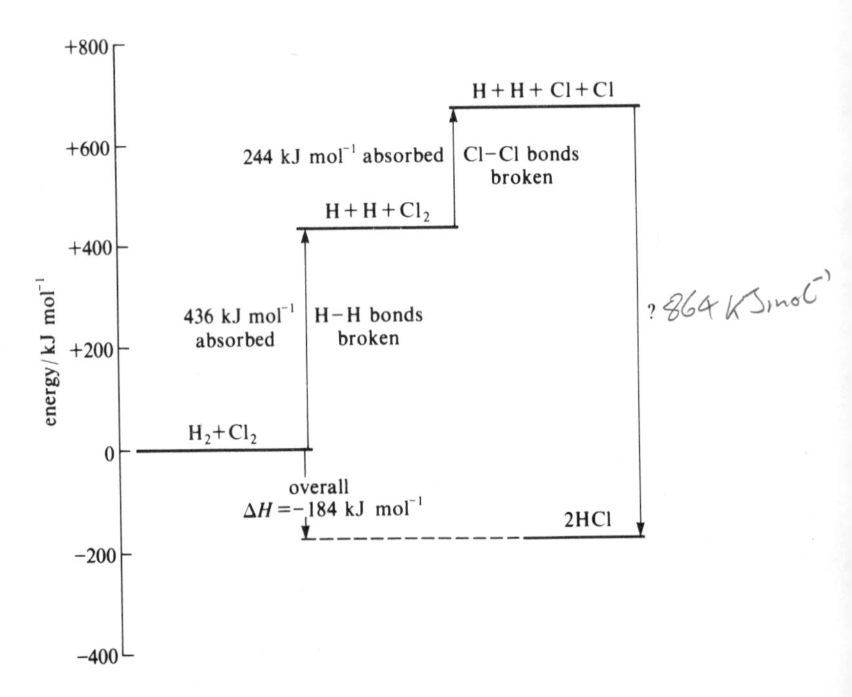

FIGURE 7 Relationship between bond energies and the overall enthalpy change of the reaction $H_2(g) + Cl_2(g) = 2HCl(g)$

Obviously it is a *bond-making* process. You have seen that bond-breaking is an endothermic process: conversely, bond-making between atoms is a downhill process in energy terms. When H atoms and Cl atoms combine to form H—Cl molecules, they achieve a state of *lower* energy. This is implied by the downward-pointing arrow that completes the overall reaction in Figure 7.

□ Look back at Table 1. What is the enthalpy change for the following process?

$$H(g) + Cl(g) = HCl(g) \tag{15}$$

■ It requires an energy equivalent to D(H—Cl), that is, 432 kJ, to break one mole of H—Cl bonds. The process in Equation 15 is exactly the reverse. According to energy conservation, this process must be exothermic, *releasing* an equivalent amount of energy. Thus,

$$\Delta H(15) = -D(\text{H—Cl}) = -432\,\text{kJ mol}^{-1}$$

☐ What then, is the enthalpy change for the process in Equation 14?

■ 432 kJ are released on formation of one mole of H—Cl bonds, so $2 \times 432\,\text{kJ} = 864\,\text{kJ}$ will be released when two moles are formed; that is

$$\Delta H(14) = -2D(\text{H—Cl}) = -864\,\text{kJ mol}^{-1}$$

Write this value (that is, $864\,\text{kJ mol}^{-1}$ released) against the right-hand arrow in Figure 7 if you wish.

Now the reaction of interest

$$H_2(g) + Cl_2(g) = 2HCl(g) \qquad (10)^*$$

is simply the *sum* of Equations 12, 13 and 14: the left-hand side, $H_2(g) + Cl_2(g)$, can be converted into the right-hand side, $2HCl(g)$, either by the direct, one-step route of Equation 10, or by the three-step route of Equations 12, 13 and 14.

Now in Unit 9, you learnt that energy can be neither created nor destroyed. What this means in this context is that the enthalpy changes for the two routes must be equal. To prove this, imagine for a moment that it was not so. Then the conversion of hydrogen and chlorine into hydrogen chloride could be carried out by one route, and then reversed along the other, finishing with the hydrogen and chlorine one started with *plus surplus energy*. As this violates the law of conservation of energy, the enthalpy changes of the two routes must be equal: the values of ΔH for Equations 12 to 14 must add up to yield the value of ΔH for the overall reaction, Equation 10. Convince yourself that this is so by doing the sum below:

$$H_2(g) = 2H(g); \quad \Delta H = D(\text{H—H}) = +436\,\text{kJ mol}^{-1}$$

$$Cl_2(g) = 2Cl(g); \quad \Delta H = D(\text{Cl—Cl}) = +244\,\text{kJ mol}^{-1}$$

$$2H(g) + 2Cl(g) = 2HCl(g); \quad \Delta H = -2D(\text{H—Cl}) = -864\,\text{kJ mol}^{-1}$$

$$H_2(g) + Cl_2(g) = 2HCl(g) \quad \Delta H = -184$$

This important idea was established empirically by W. G. Hess in 1840. It is therefore called **Hess's law**, and may be stated as follows:

An energy change for a chemical reaction is the same whether the reaction takes place in just one step, or by a number of separate steps whose sum is equal to the one-step process.

What you have just read should have convinced you that Hess's law is just a special case of the law of conservation of energy.

☐ Look again at Figure 7. How can the overall enthalpy change for Reaction 10 be related to the bond energies of the individual bonds that are broken and formed?

■ The overall enthalpy change is the *difference* between the energy required to break the H—H and Cl—Cl bonds, and that released when the H—Cl bonds are formed:

$$\Delta H(10) = D(\text{H—H}) + D(\text{Cl—Cl}) - 2D(\text{H—Cl})$$

Try the following ITQ to check your understanding of this method for calculating enthalpy changes.

ITQ 4 Use appropriate bond energies from Table 1 to calculate the enthalpy change for the following reaction:

$$H_2(g) + F_2(g) = 2HF(g)$$

+436
+158 = 594 kJ mol⁻¹ − 2×568
= −542 kJ mol⁻¹

Does your calculation suggest a possible reason why this reaction is so exothermic?

because of strong bonds H-H & weak bonds F-F combining into strong exothermic reaction.

AVERAGE BOND ENERGY

These calculations incorporate a very important idea. When calculating enthalpy changes, it is not necessary to know whether or not, for example, the reaction in Equation 10 actually takes place as described above, that is, hydrogen and chlorine molecules first decomposing, and then the H and Cl atoms combining to form HCl molecules. Indeed, you will see later (Section 5) that this particular path is extremely unlikely. But here this is unimportant, because the *overall* energy change does *not* depend on the path actually taken.

So far, the discussion has concentrated on diatomic molecules, in which the bond energy is simply the energy required to dissociate the molecule. Each bond in a *polyatomic* molecule (NH_3 or CH_4, for example) also has a characteristic bond energy associated with it, but in this case the energy is not quite as easy to pin down. The problem with polyatomic molecules is that the bond energy depends not only on the types of atom connected by the bond, and whether the bond is single, double, etc.; it also depends on the nature of the rest of the molecule, that is, on the bond's environment. Consider methane, CH_4, for example.

□ Write a Lewis structure for CH_4.

■ As you saw in Units 13–14, carbon has four electrons in its outermost shell, whereas hydrogen has one. In CH_4, all the atoms can attain noble gas configurations by sharing pairs of electrons, as:

```
     H                  H
     × •                |
  H •× C ו H    or   H—C—H
     • ×                |
     H                  H
```

Thus, CH_4 contains four carbon–hydrogen bonds. The energies required to break successive C—H bonds in CH_4 are shown in Table 2: evidently, they are not all the same. However, you have seen how useful bond energies are in calculating ΔH for a reaction from the energies involved in breaking and making bonds. For this purpose, collections of **average bond energies** (for example, the average value in Table 2) are available, giving representative bond energies based on a large selection of compounds in which that bond appears. A few values of these are given in Table 3.

TABLE 2 Carbon–hydrogen bond energies in CH_4 at 25 °C

Bond	D/kJ mol^{-1}
$H_3C—H = H_3C + H$	426
$H_2C—H = H_2C + H$	481
$HC—H = HC + H$	417
$C—H = C + H$	339
	average $\frac{1\,663}{4} \approx 416$

TABLE 3 Some average bond energies at 25 °C

Bond	Average bond energy/kJ mol^{-1}
C—C	330
C=C	589
C≡C	811
C—H	416
C—Cl	326
C—O	327
C=O	804
O—O	143
O—H	463
N—N	159
N—H	391

Notice that many of the bonds listed in Table 3 have carbon as one of the atoms involved. These are the kinds of bond present in the carbon compounds that you will study in Units 17–18. Several of these bonds will be referred to again during the discussion of chemical fuels in Section 4. Two of the other values listed in Table 3 are also worth a comment, namely the bond energies for the oxygen–oxygen (O—O) and nitrogen–nitrogen (N—N) *single* bonds. These bonds are present in the compounds hydrogen peroxide (H_2O_2) and hydrazine (N_2H_4), respectively. Convince yourself that this is so by trying the following ITQ.

ITQ 5 Write Lewis structures for H_2O_2 and N_2H_4. Assume that the two oxygen or two nitrogen atoms are bonded to each other and that an equal number of H atoms are bound to each O or N.

The average bond energies for O—O and N—N can be compared with those for the corresponding multiple bonds in the diatomic molecules: from Table 1, $D(\text{O=O}) = 498\,\text{kJ mol}^{-1}$ and $D(\text{N}\equiv\text{N}) = 945\,\text{kJ mol}^{-1}$. In each case, the single bond is considerably weaker, and this has important consequences for the chemistry of compounds containing these bonds. The TV programme 'Energy and rockets' shows an example of this.

Average bond energies such as those in Table 3 can be used to calculate enthalpy changes in just the same way as before. Since they are average values, however, the resulting values of ΔH are only approximate. Nevertheless, they give a good indication of the magnitude of the energy changes involved. Consequently, in this Course, we shall often combine the strict bond energies of Table 1 with the average bond energies of Table 3 in the same calculation, using the symbol D for both.

For example consider the following reaction:

$$3H_2(g) + N_2(g) = 2NH_3(g) \qquad (16)$$

As you found in ITQ 3, this reaction involves breaking three H—H bonds and one N≡N bond for every six N—H bonds formed. Thus

$$\Delta H = 3D(\text{H—H}) + D(\text{N}\equiv\text{N}) - 6D(\text{N—H})$$

The required bond energies are given in Tables 1 and 3, so

$$\Delta H = (3 \times 436 + 945 - 6 \times 391)\,\text{kJ mol}^{-1}$$

$$= -93\,\text{kJ mol}^{-1}$$

ITQ 6 List the bonds broken and formed in the following reaction:

$$2H_2(g) + O_2(g) = 2H_2O(g)$$

Start by drawing a Lewis structure for H_2O.

Using appropriate values from Tables 1 and 3, calculate the enthalpy change for this reaction.

3.3 LIMITATIONS ON THE USE OF BOND ENERGIES

The example in ITQ 6 highlights an important point about the use of bond energies: the calculation you performed refers to the following reaction:

$$2H_2(g) + O_2(g) = 2H_2O(g); \quad \Delta H = -482\,\text{kJ mol}^{-1} \qquad (17)$$

in which *gaseous* water is formed. Suppose now that you wanted to calculate ΔH for the analogous reaction in which liquid water is formed:

$$2H_2(g) + O_2(g) = 2H_2O(l) \qquad (18)$$

In Section 2, you met the following process:

$$H_2O(l) = H_2O(g); \quad \Delta H = +43.3\,\text{kJ mol}^{-1} \qquad (9)^*$$

□ What is the enthalpy change for the reverse process, the condensation of gaseous water to liquid?

$$H_2O(g) = H_2O(l) \qquad (19)$$

■ By the law of conservation of energy, this must be an exothermic process, releasing an equivalent amount of energy (cf SAQs 2 and 3), so

$$\Delta H(19) = -43.3\,\text{kJ mol}^{-1}$$

Can you see the connection between Equations 17, 18 and 19?

Equation 18 is the sum of Equation 17 and twice Equation 19. Remember that if an equation is doubled, then the value of ΔH must also be doubled:

$$2H_2O(g) = 2H_2O(l); \quad \Delta H = 2 \times -43.3 = -86.6\,\text{kJ mol}^{-1} \qquad (20)$$

According to Hess's law, the values of ΔH for Equation 17 and 20 must also add to give that for Equation 18:

$$2H_2(g) + O_2(g) = 2H_2O(g); \quad \Delta H = -482\,\text{kJ mol}^{-1} \qquad (17)^*$$

$$2H_2O(g) = 2H_2O(l); \quad \Delta H = -86.6\,\text{kJ mol}^{-1} \qquad (20)^*$$

$$2H_2(g) + O_2(g) = 2H_2O(l); \quad \Delta H = -568.6\,\text{kJ mol}^{-1} \qquad (18)^*$$

This example shows quite clearly that the use of bond energies gives a reasonable estimate of ΔH only when all the reactants and products are gaseous species held together by simple covalent bonds. *Bond energies must be used carefully.* If the reaction of interest involves liquids (or solids) as well as gases, then this must be taken into account by including in the calculation values of ΔH for appropriate phase changes.

SUMMARY OF SECTION 3

1 The energy changes associated with simple gaseous reactions can be interpreted in terms of changes in the way the atoms are bound together.

2 Breaking a covalent bond requires an input of energy equivalent to the bond energy. Conversely, when atoms combine to form a bond, an equivalent amount of energy is released.

3 When calculating energy changes for chemical reactions, Hess's law is invaluable: it states that the energy change is the same whether the reaction takes place in just one step, or by a number of separate steps whose sum is equal to the one-step process.

4 Bond energies can be used to calculate the overall enthalpy change for a simple gas reaction involving either diatomic or polyatomic molecules (or both).

5 If the reaction of interest involves liquids (or solids) as well as gases, then values of ΔH for appropriate phase changes must be included in the calculation.

SAQ 4 This question leads in to the study of chemical fuels in the next Section.

(a) Write a Lewis structure for carbon dioxide, CO_2. Then list the bonds broken and formed in the following reaction:

$$C(g) + O_2(g) = CO_2(g)$$

(b) Using appropriate bond energies from Tables 1 and 3, calculate the enthalpy change for the reaction above. Compare your answer with the value, $-393.5\,\text{kJ mol}^{-1}$, of ΔH for Reaction 1 in Section 1.

How do you account for any discrepancy between the two values?

SAQ 5 The decomposition of hydrogen peroxide, H_2O_2, is an exothermic process. The energy released has been used for propulsion in space walks.

(a) Consider the decomposition of gaseous H_2O_2:

$$2H_2O_2(g) = 2H_2O(g) + O_2(g)$$

Using appropriate bond energies from Tables 1 and 3, calculate ΔH for this reaction. (Refer back to the Lewis structures for H_2O_2 and H_2O, which you used to answer ITQs 5 and 6.)

Does your calculation suggest why this reaction is so exothermic?

(b) Given the following enthalpy changes:

$$H_2O(l) = H_2O(g); \quad \Delta H = +43.3\,\text{kJ mol}^{-1}$$

$$H_2O_2(l) = H_2O_2(g); \quad \Delta H = +51.6\,\text{kJ mol}^{-1}$$

calculate ΔH for the decomposition of liquid H_2O_2:

$$2H_2O_2(l) = 2H_2O(l) + O_2(g)$$

SAQ 6 Explain briefly why it is not possible to calculate ΔH for the following reaction from simple bond energy data:

$$HCl(g) = H^+(aq) + Cl^-(aq)$$

4 CHEMICAL FUELS

As we said in the Introduction, most of the energy 'used' today comes originally from fossil fuels: coal, oil and natural gas. The characteristic property of such fuels lies in their reaction with oxygen in the air: energy is released when they are burnt in oxygen. So the concept of fuels and their energies really has to do with a 'fuel package', molecules to be burned, oxygen molecules 'to do the burning': a reshuffling of matter takes place at the molecular level, product molecules are formed and energy is released. For most conventional uses, chemical energy is first released as heat, which is either used directly or subject to several transformations. On purely chemical grounds, one of the main criteria of a 'good' fuel is the amount of heat that can be obtained when it is burnt, that is, the magnitude of ΔH for the combustion reaction.

Table 4 contains information on the combustion reactions of typical chemical fuels which are used in the internal combustion engine, in industry and the home, and in our bodies. It is evident from Table 4 that these fuels all burn in oxygen to yield the same two products, carbon dioxide and water. In order to compare the fuels, it has been assumed that the water formed is in the gaseous state, although this will not be the case for reactions occurring in the body. We shall now use the ideas developed in the previous Section to examine why such reactions are exothermic.

Let us start by considering one of the simplest examples in Table 4, the burning of natural gas (mostly methane), and try to interpret the overall energy change in terms of the bonds broken and formed.

$$CH_4(g) + 2O_2(g) = CO_2(g) + 2H_2O(g) \qquad (2)^*$$

☐ List the bonds broken and formed when Reaction 2 takes place.

TABLE 4 Combustion of typical fuels in oxygen*

Fuel and combustion reaction	ΔH / kJ per mol fuel	ΔH / kJ per g fuel
carbon (coke) $C(s) + O_2(g) = CO_2(g)$	-394	-33
methane (natural gas) $CH_4(g) + 2O_2(g) = CO_2(g) + 2H_2O(g)$	-800	-50
octane (petrol)† $C_8H_{18}(l) + 12\frac{1}{2}O_2(g) = 8CO_2(g) + 9H_2O(g)$	$-5\,060$	-44
stearic acid (in animal fats) $C_{18}H_{36}O_2(s) + 26O_2(g) = 18CO_2(g) + 18H_2O(g)$	$-10\,570$	-37
glucose (in the body) $C_6H_{12}O_6(s) + 6O_2(g) = 6CO_2(g) + 6H_2O(g)$	$-2\,560$	-14
hydrogen $H_2(g) + \frac{1}{2}O_2(g) = H_2O(g)$	-241	-120

* The values given in this Table refer to combustion of the stated amount of *pure* fuel, for example, pure methane. The actual amount of heat that can be obtained from a given sample of petrol, natural gas, etc., depends very much on its chemical purity.

† Octane is just one of the many components of commercial petrol.

■ Recalling the structures of CH_4, O_2, CO_2 and H_2O, four C—H and two O=O bonds are broken for every four O—H and two C=O bonds formed.

Gathering together information from Tables 1 and 3 gives:

	Single	D/kJ mol^{-1}	Double	D/kJ mol^{-1}
Broken	4 C—H	4 × 416	2 O=O	2 × 498
Made	4 O—H	4 × 463	2 C=O	2 × 804

□ Use these values to check the value of ΔH for Reaction 2 given in Table 4.

■ $$\Delta H = (4 \times 416 + 2 \times 498 - 4 \times 463 - 2 \times 804)\,\text{kJ mol}^{-1}$$
$$= -800\,\text{kJ mol}^{-1}$$

Notice that in this reaction the same numbers of single and double bonds are broken and made. However, although C—H and O—H bonds are of comparable strength, the C=O bond is over one and a half times as strong as the O=O bond.

Does this suggest why combustion of CH_4 is an exothermic reaction?

It seems that the exothermic nature of this reaction can be largely attributed to the extreme strength of the C=O bonds in CO_2: considerable energy is released when these bonds are formed. It is often said that energy is 'stored' in natural gas. According to the molecular picture outlined above, a better description would be 'the fuel package (methane plus oxygen) is a source of energy because the combustion process yields a product with *stronger* bonds (higher bond energy).'

But what about the other combustion reactions in Table 4, excepting hydrogen for the moment?

Look back at Table 4. Can you see a problem with this simple analysis in terms of bond energies?

In the remaining reactions, the fuel is either a liquid or a solid. However, the phase changes from gas to liquid and from gas to solid are always exothermic (cf. the discussion in Sections 2 and 3.3), so hypothetical reactions involving *gaseous* fuels would in all cases be even more exothermic. If you are unsure about this, look back at the answers and comments to SAQs 4 and 5.

At first sight there seems to be another problem with the simple analysis outlined above: the numbers of bonds broken and made are not necessarily the same. Consider octane as an example. Like methane, octane is a member of a large class of compounds known as *hydrocarbons*. You will study the structures and properties of these compounds in detail in Units 17–18. For the moment, we ask you to accept that octane contains seven C—C bonds and 18 C—H bonds.

□ With reference to the octane combustion reaction in Table 4, how many bonds are broken and made when gaseous octane is burned?

■ Seven C—C, 18 C—H and $12\frac{1}{2}$ O=O bonds are broken for every 16 C=O and 18 O—H bonds made.

Thus, 34 bonds are made for every $37\frac{1}{2}$ that are broken (both totals including a number of multiple bonds). In other words, more bonds are broken than made, but the process is *still* highly exothermic. However, referring to the values in Table 3, you can see that both C—H ($D = 416\,\text{kJ mol}^{-1}$) bonds and C—C ($D = 330\,\text{kJ mol}^{-1}$) bonds are considerably weaker than the C=O bonds in CO_2 ($D = 804\,\text{kJ mol}^{-1}$). Again, it seems reasonable to attribute the exothermic nature of this reaction to the extreme strength of the $C{=}O$ bonds.

The structures of stearic acid and glucose are too complex to consider here. However, the combustion of these, and other *carbon* compounds in oxygen also yields carbon dioxide. It seems reasonable to conclude that the exothermic nature of such reactions results mainly from making the very strong $C{=}O$ bonds at the expense of breaking weaker ones.

Notice that this argument does not apply to the combustion of hydrogen: $H{-}H$, $O{=}O$ and $O{-}H$ bonds are of comparable strength. In this case the decisive factor is the relative number of bonds broken (one $H{-}H$ and a half $O{=}O$) and made (two $O{-}H$).

As a final point, consider the figures in the last column of Table 4. These were obtained by dividing the figures in the preceding column by the molar mass of the fuel in the given chemical equation. It is immediately apparent that on a weight-for-weight basis, hydrogen is by far the 'best' fuel, by a factor of three over petrol. This, together with the growing shortage of conventional fossil fuels and their associated pollution problems, has led to a considerable interest in the development of hydrogen as a synthetic chemical fuel. The photograph in Figure 8 suggests that at the time of writing, hydrogen-power is a very real possibility for the future.

FIGURE 8 The world's first hydrogen-powered mass transit vehicle, developed by the Billings Corporation of Provo, Utah, has demonstrated the feasibility and practicability of hydrogen as a fuel. This pollution-free, prototype bus safely stores hydrogen in small iron–titanium particles, which when heated supply hydrogen to the easily modified diesel engine.

The fuels discussed in this Section are all burnt in *oxygen* in order to release energy. The reason for using oxygen as the so-called 'oxidizer' is not hard to find: it is freely available in the air around us. But you have seen earlier in this Unit that combustion in oxygen is not the only type of exothermic reaction. Indeed, the simple analysis outlined in this and the previous Section suggests that any gaseous reaction will be exothermic if it results in the formation of strong bonds from weaker bonds. For example, as you found in answering ITQ 4, the weakness of the $F{-}F$ bond explains why the reactions of many substances with fluorine are strongly exothermic.

However, fluorine is a thoroughly nasty material, and nobody in their right mind would suggest replacing oxygen with fluorine as the oxidizer for conventional uses. On the other hand, exotic, and often unpleasant, oxidizers and fuels are used in more specialized areas. One such area, the propellant systems used to power rockets, is examined in the TV programme for this Unit.

REACTION MECHANISM

SUMMARY OF SECTION 4

1 The energy obtained from most chemical fuels is the energy released in a reaction with oxygen.

2 An analysis using average bond energies suggests that the energy obtained from hydrocarbon fuels is high mainly because the C=O bond is much stronger than the O=O bond.

3 The combustion of a given mass of hydrogen releases much more energy than does the combustion of the same mass of other common fuels.

SAQ 7 As you will see in the TV programme 'Energy and rockets', the reaction between hydrogen peroxide (the oxidizer) and hydrazine (the fuel) was used to power the Messerschmidt 163.

Calculate ΔH for the reaction

$$2H_2O_2(g) + N_2H_4(g) = 4H_2O(g) + N_2(g)$$

(Refer back to ITQ 5 for the structures of H_2O_2 and N_2H_4.) Why is this reaction so exothermic?

5 RATES OF CHEMICAL REACTIONS

As the information in Table 4 showed, the combustion reactions discussed in the previous Section are all strongly exothermic. In addition, these highly exothermic reactions have equilibrium constants with truly enormous values at normal temperatures. For example, at 25 °C, the combustion of methane

$$CH_4(g) + 2O_2(g) = CO_2(g) + 2H_2O(g) \qquad (2)^*$$

has an equilibrium constant of about 10^{140}—an enormous number!

□ What does this high value of K mean? (Refer back to Section 7 of Unit 15 if necessary.)

■ The high value of K implies that *at equilibrium* reactants should be effectively completely converted into products.

Now oxygen is readily available from the atmosphere, and (leaving aside the water content of living organisms) we and all living things are composed of carbon compounds. Why then does not everything that is potentially flammable immediately burst into flame, including ourselves? As implied in the Introduction, the answer is that all such reactions are incredibly *slow* under normal conditions. They effectively never reach the very favourable equilibrium position.* Lucky for us!

In a similar way, the value of K for the reaction between gaseous hydrogen and chlorine (analogous to that between hydrogen and iodine, cf Unit 15 Sections 1 and 7):

$$H_2(g) + Cl_2(g) = 2HCl(g); \quad K = 2.5 \times 10^{33} \text{ at } 25\,°C \qquad (10)^*$$

implies that the *equilibrium* position strongly favours formation of the product, hydrogen chloride, under normal conditions. However, the size of K contains no hint of the experimental observation that a mixture of

* Referring again to Section 1, this is why you have to heat methane in order to fire it. This speeds up the reaction. Notice, however, that according to the discussion in Section 9 of Unit 15, changing the temperature will also change the value of K: in this case, K actually decreases with increasing temperature (it's about 10^{32} at 1000 °C). This point will be taken up in Sections 6 and 7.

hydrogen and chlorine can be left for several days in a dark room with no noticeable reaction taking place, but that the mixture explodes violently when exposed *briefly* to a bright light.

It is fairly obvious that this aspect of chemical reactions is of crucial importance in the chemical industry. As you saw in Unit 15, the value of K for a reaction indicates the *maximum* yield of product that can be achieved, in principle, under a given set of conditions, but the examples cited above suggest that it says nothing about *how fast* this equilibrium position will be attained. From the equilibrium constant alone it is impossible to tell whether the reaction will take an inconveniently long time to reach equilibrium or whether it will proceed with explosive violence. It is evidently important to know not only how fast a reaction will go, but also how this speed or *rate* may be influenced or controlled by changing the reaction conditions.

However, chemists have other interests in reaction rates apart from the problems associated with the chemical industry. In particular, the study of reaction rates provides one of the most powerful methods for testing models of what actually happens on the *molecular* level during a reaction: this is called the **reaction mechanism**. The theme to be developed here is that, unlike the equilibrium constant and the overall energy change for a reaction, the *rate* of a reaction depends on the *route* that it takes—on its mechanism.

5.1 WHAT INFLUENCES THE RATE OF A CHEMICAL REACTION?

Two important ways of changing the rate of a chemical reaction can be demonstrated by mixing aqueous solutions of potassium persulphate, $K_2S_2O_8$, and potassium iodide, KI. In water $K_2S_2O_8$ forms the persulphate ion, $S_2O_8^{2-}$(aq):

$$K_2S_2O_8(s) = 2K^+(aq) + S_2O_8^{2-}(aq) \quad (21)$$

and KI forms the iodide ion, I^-(aq):

$$KI(s) = K^+(aq) + I^-(aq) \quad (22)$$

Suppose that suitable aqueous solutions of these substances are quickly mixed to give one litre of a mixture which would, if no reaction occurred, contain 0.1 mole of KI and 0.01 mole of $K_2S_2O_8$. Now we have chosen these amounts because they give a convenient reaction rate: a reaction *does* occur, and the initially colourless mixture gradually turns brown. The persulphate and iodide ions gradually react together to form sulphate ions and iodine:

$$S_2O_8^{2-}(aq) + 2I^-(aq) = 2SO_4^{2-}(aq) + I_2(aq)$$

At 25 °C, with the concentrations specified above, the amount of iodine in the solution reaches 0.005 mole after about $1\frac{3}{4}$ minutes. At this moment exactly half of the iodine that would be produced by the *complete* reaction of the 0.01 moles of persulphate has been formed.

Suppose now that the reaction is repeated with the same mixture, but the temperature is 35 °C rather than 25 °C. This time, the 0.005 mole of I_2 appears after only about $\frac{3}{4}$ minute. A very similar result can be obtained in a different way: by doubling the initial concentration of persulphate. As before, the litre of mixture contains 0.1 mole of KI, but the amount of dissolved potassium persulphate, and therefore of $S_2O_8^{2-}$(aq), is doubled to 0.02 mole. This time, 0.005 mole of I_2 is again formed after only about $\frac{3}{4}$ minute, even though the reaction temperature is only 25 °C.

These results illustrate something of great general importance. *The rate of a chemical reaction can often be increased by increasing the concentration of the reactants, or by raising the reaction temperature.* Why should this be so?

COLLISION MODEL OF A CHEMICAL REACTION

ACTIVATION ENERGY

5.2 MOLECULAR INTERPRETATION OF THE EFFECT OF CONCENTRATION AND TEMPERATURE

Why should the rate of a reaction depend on the concentrations of the reactants, usually increasing as they are increased?

□ What happens at the molecular level when the concentration of a substance is increased?

■ Increasing the concentration increases the number of molecules in a given volume.

Intuitively, it seems reasonable to suggest that two molecules must meet or 'collide' before they react. According to this **collision model of a chemical reaction**, the rate of a reaction should be related to the frequency with which molecules collide—the *collision frequency*.

□ If the number of molecules in a given volume is increased, would you expect the collision frequency to increase or to decrease?

■ With more molecules in a given volume, you would expect collisions to occur more often, that is, the collision frequency should increase.

If the rate of a reaction increases with the collision frequency, then increasing the concentration of a reactant should increase the reaction rate. In other words, the collision idea fits our experimental findings in a qualitative way. To examine this collision model a bit more closely, we shall restrict the discussion to the simplest kind of process—gas reactions. As with the discussion of energy changes in Section 3, when a reaction occurs in solution, the presence of solvent molecules complicates the picture.

Consider the following particularly simple example, the reaction between bromine atoms and hydrogen molecules to produce hydrogen bromide molecules and hydrogen atoms:

$$Br(g) + H_2(g) \longrightarrow HBr(g) + H(g) \qquad (23)$$

Imagine a bromine atom moving around in a vessel containing hydrogen molecules. What factors will be important in determining how often it gets close enough to an H_2 molecule to collide? One factor is obviously the number of H_2 molecules in a given volume, that is, the concentration of hydrogen: this factor was discussed above. However, there is another factor, the relative speed of the bromine atom and the hydrogen molecules: the faster they move relative to one another, the more frequently they will collide. According to a collision model therefore, the rate of Reaction 23 will increase if this relative speed is increased.

Such an increase in relative molecular speed is just what is achieved by raising the temperature. From Unit 9 you know that the temperature of a gas rises when heat is transferred to it. The increase in energy brought about by heat transfer is shared among the individual molecules of the gas, principally in the form of kinetic energy.

□ What is the relationship between the mass m of a molecule, its speed, v, and its kinetic energy E_k?

■ $E_k = \frac{1}{2}mv^2$ (Unit 9).

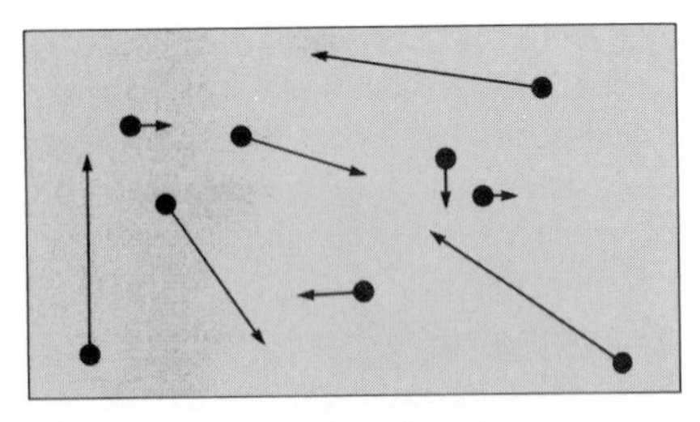

FIGURE 9 Atoms moving around randomly and independently in a gas.

It follows that if the kinetic energy of a molecule increases, its speed, v, must increase: the net effect of increasing the temperature of a gas is to increase the *average* speed of the molecules. We stress the word *average*, because the molecules in a gas move at different speeds. Figure 9 represents the molecular model of a gas used in this Unit. The molecules move randomly with respect to one another. The arrows attached to the molecules mark directions of motion; their lengths are a measure of molecular speed. At any instant, the speed of some molecules is less than average; the speed of others is greater. Moreover, kinetic energy is constantly being exchanged between molecules by collisions, so that a molecule moving quickly at one instant may be moving slowly at the next. Nevertheless, at any temperature, there will be a definite *average* speed, and this overall average is increased by a temperature rise.

In summary, the average speed of the molecules in a gas is greater at higher temperatures. In terms of the collision model, the collision frequency and hence the rate of reaction should increase with increasing temperature. Again this predicted behaviour agrees qualitatively with experimental findings: all gas reactions do go faster as the temperature is increased.

So far so good: in qualitative terms, the simple collision model predicts a dependence of reaction rates on concentration and on temperature which agrees in kind with the observed behaviour. However, there are problems with this simple model, which become apparent as soon as a more quantitative comparison is made between theory and experiment. By treating molecules as tiny spheres with a characteristic size and assuming that they travel at some *average* speed at a given temperature, it is possible to calculate the rate of collision as a function of concentration, temperature, etc. When such calculations are compared with the observed rate of a given reaction, important discrepancies become apparent. In particular, at a fixed temperature, the calculated rate of collision between molecules is often considerably higher than the observed rate of reaction, by a factor of 10^8 for Reaction 23 for example! It seems that the collision frequency provides an upper limit, and that, in general, most collisions do not lead to reaction.

Why should the reaction rate be lower than the collision frequency? Well, when two molecules collide, they may *either* react to form product molecules *or* they may rebound and go harmlessly on their way: in other words, not all collisions are equally effective in causing reaction. Why should this be so? A possible explanation is that molecules will react only if they collide with an energy equal to or greater than a certain minimum value. A gentle collision does not result in reaction; a violent collision does. Perhaps only those molecules that collide with an energy much larger than average can actually react. This suggests that there may be a *threshold energy* required to permit reaction, and that a collision leads to reaction only when the energy associated with it is equal to or greater than this threshold value.

This possible addition to the simple collision picture of chemical reaction would certainly provide a plausible explanation of why a proportion of collisions do not lead to reaction. Let us examine the idea in more detail.

5.3 THE ACTIVATION ENERGY

If most molecules do not possess enough energy to react, it must be because the energy threshold, known as the **activation energy** and given the symbol E_a, is greater than the average energy of a collision. However, from Section 5.2, you know that when the temperature is raised, the average kinetic energy of the gas molecules is increased. This in turn means that the proportion of collisions with an energy above the activation energy increases with temperature. Thus our modified collision model still predicts that the outcome of an increase in temperature is an increase in the rate of reaction.

This argument certainly agrees with the observed experimental behaviour. Moreover, although we shall not discuss it in detail here, by studying the variation in reaction rate with temperature, experimental values for the activation energies of reaction can be determined. For Reaction 23, for example, E_a is found to be about $82\,\text{kJ}\,\text{mol}^{-1}$.

REACTION-COORDINATE DIAGRAM

ACTIVATED COMPLEX

But does this picture of a threshold energy for reaction make sense? A good analogy is provided by the collision of one car with another:*

> If (the cars) are both moving slowly, they'll bounce off each other and there will be no more damage than, possibly, heated emotions. If one of the cars is moving quite rapidly, however, there may be a significant rearrangement of the parts of each car. A bumper may be knocked off, the door from car A may leave the scene embedded in the hood of car B, and so on. It takes some minimum amount of energy though to break off a bumper and some different energy to wrench a door loose from its hinges. These would be the threshold energies for those particular automotive changes.

Similarly, for simple gas reactions, the reactant molecules are held together with covalent bonds: these must be broken, at least partially, before the atoms can be rearranged into product molecules—a process which you have seen requires energy. To examine this idea a bit more closely, consider again the very simple gas reaction

$$Br(g) + H_2(g) \longrightarrow HBr(g) + H(g) \qquad (23)^*$$

□ What is the *overall* enthalpy change for Reaction 23? Is the reaction endothermic or exothermic?

■ Overall, H—H bonds are broken and an equal number of H—Br bonds are formed. (Remember that you do not need to know the energies of the H and Br atoms: only changes in bonding are important.) Using bond energies from Table 1:

$$\Delta H = D(\text{H—H}) - D(\text{H—Br})$$
$$= (436 - 366)\,\text{kJ mol}^{-1}$$
$$= 70\,\text{kJ mol}^{-1}$$

The enthalpy change is positive, so the reaction is endothermic.

Recalling Figure 4b (Section 2), this overall energy change can be represented as shown in Figure 10: the enthalpy change for the reaction is the difference between the energy of the reactants and that of the products. But how does the energy change *during the course* of the reaction, that is, in the central part of the diagram in Figure 10?

Well, the 'reactant' level on the left-hand side of Figure 10 represents the situation when a bromine atom and a hydrogen molecule have a large separation: they do not influence one another. However, when a bromine atom gets quite close to a hydrogen molecule, it enters a region where repulsive interactions are strong enough to oppose any closer approach unless the system has enough energy to overcome these forces.

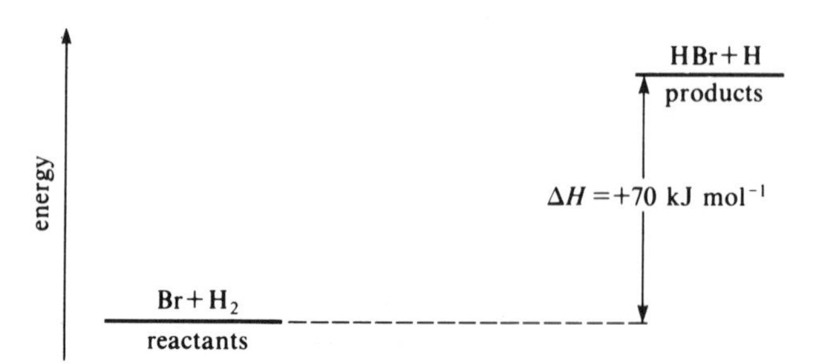

FIGURE 10 Energy diagram for the reaction $Br(g) + H_2(g) \longrightarrow HBr(g) + H(g)$.

□ Can you think of any repulsive interactions that might be present?

■ In simple electrostatic terms, there will be repulsion between the negatively charged electrons of the Br atom and those of the H_2 molecule (and between the Br and H nuclei). Remember that charges of like sign repel one another (Unit 9).

* G. C. Pimentel and R. D. Spratley (1971) *Understanding Chemistry*, Holden-Day.

At the same time, the electrons of the H_2 molecule will start to be attracted by the bromine nucleus, and this will weaken the H—H bond. This effect will continue until the system passes through an energy *maximum*, beyond which the bonding between Br and H increases, whereas that between H and H decreases to the extent that the situation is better described as an HBr molecule and an H atom: this situation is represented by the 'product' level on the right-hand side of Figure 10. Chemists like to represent this change pictorially by filling in the centre of the diagram as shown in Figure 11. Diagrams like this are called **reaction-coordinate diagrams**, because the horizontal axis is often referred to as the 'reaction coordinate', a term that is used to represent the progress of the reaction. Movement along the reaction coordinate from left to right is equivalent to those gradual changes that occur in the distances between the bromine atom and the two hydrogen atoms as the reactants change into the products.

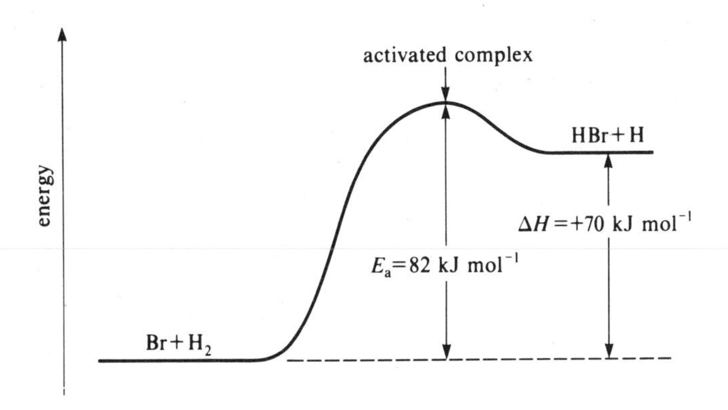

FIGURE 11 Reaction-coordinate diagram for the reaction

$Br(g) + H_2(g) \longrightarrow HBr(g) + H(g)$

The state of the system when the energy is a maximum corresponds to a hypothetical arrangement in which reactant bonds are partially broken and product bonds are partially formed, an arrangement that is often called the **activated complex**. For this particular reaction, the structure can be represented as follows

Br---H---H

where the broken lines imply partial bonding (fewer than two electrons shared between two atoms): the position of the activated complex is shown on Figure 11. In general, however, it is not possible to predict the exact structure of the activated complex, other than that it represents a transition between reactants and products.

According to this pictorial representation, the difference in energy between the reactants and the activated complex is the activation energy for the reaction, also shown on Figure 11. Thus, E_a can be thought of as an *energy barrier* that reactant molecules must surmount before products can be formed. For reaction to take place, a bromine atom must collide with a hydrogen molecule and the joint system must have energy E_a more than the average energy of Br and H_2; otherwise it cannot cross the energy barrier.

Not only does Figure 11 give a pictorial view of the activation energy as an energy barrier to reaction, but it also illustrates an extremely important point, which was mentioned frequently in Section 3: the overall energy change, ΔH, does *not* depend on the size of the energy barrier, that is, on the path the reaction takes.

The example discussed so far, Equation 23, is an endothermic reaction. For simple endothermic processes like this, the reaction-coordinate diagram will always have the general shape shown in Figure 11, although the size of ΔH and of the energy barrier will, of course, vary from reaction to reaction. But what about exothermic reactions?

The first example to hand is the reverse of Equation 23:

$$H(g) + HBr(g) \longrightarrow H_2(g) + Br(g) \qquad (24)$$

☐ What is the value of ΔH for Reaction 24?

CATALYST

■ By the law of conservation of energy, the reaction must be exothermic, releasing an equivalent amount of energy, so

$$\Delta H(24) = -70\,\mathrm{kJ\,mol^{-1}}$$

Now, according to the treatment in this Unit, the activation energy of a reaction must be positive.

□ Before reading further, try to sketch a reaction-coordinate diagram showing the energy changes during the course of Reaction 24. Mark on your diagram the activation energy, E_a, the overall energy change, ΔH, and the position of the activated complex.

■ As ΔH is negative, the products must be of lower energy than the reactants. The activation energy is the difference in energy between the reactants and the activated complex: if E_a is positive, there must again be a 'hump' in the middle of the diagram. A sketch is shown in Figure 12.

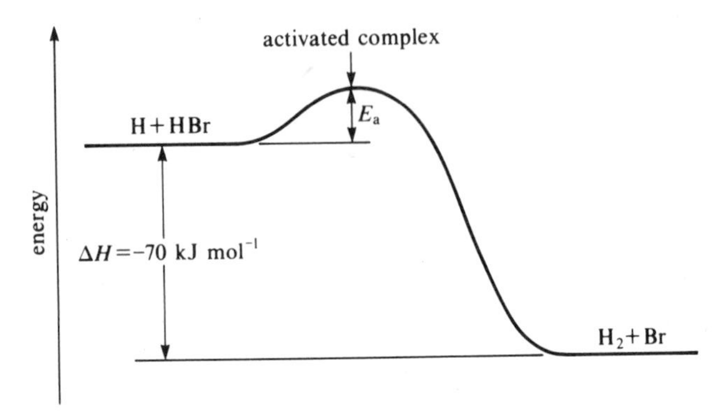

FIGURE 12 Reaction-coordinate diagram for the reaction

$H(g) + HBr(g) \longrightarrow H_2(g) + Br(g)$.

Again, any simple exothermic reaction will have a reaction-coordinate diagram similar to that in Figure 12. Notice that the general shape of Figure 12 is the reverse of the diagram for an endothermic process.

However, for the simple examples discussed in this Section it is possible to go one step further. Consider the two reactions together:

$$\text{forward reaction: } H_2(g) + Br(g) \longrightarrow H(g) + HBr(g) \qquad (23)^*$$

$$\text{reverse reaction: } H(g) + HBr(g) \longrightarrow H_2(g) + Br(g) \qquad (24)^*$$

If the path taken by the forward reaction is the same as that taken by the reverse reaction, then it is possible to combine the diagrams in Figures 11 and 12 as shown in Figure 13. In this diagram $E_a(23)$ and $E_a(24)$ are the activation energies for the forward and reverse reactions, respectively.

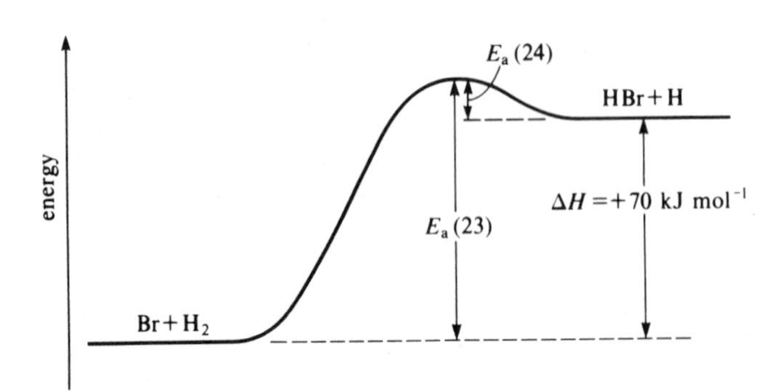

FIGURE 13 Reaction-coordinate diagram for the reaction

$Br(g) + H_2(g) = HBr(g) + H(g)$

$E_a(23)$ and $E_a(24)$ are the activation energies for the forward and reverse reactions, respectively.

□ Recalling that $E_a(23)$ is about $82\,\mathrm{kJ\,mol^{-1}}$, can you see how Figure 13 enables you to calculate $E_a(24)$?

■ It should be clear from Figure 13 that ΔH for the forward reaction is given by the difference:

$$\Delta H(23) = E_a(23) - E_a(24)$$

Thus

$$\begin{aligned} E_a(24) &= E_a(23) - \Delta H(23) \\ &= (82 - 70)\,\mathrm{kJ\,mol^{-1}} \\ &= 12\,\mathrm{kJ\,mol^{-1}} \end{aligned}$$

Provided that the same energy diagram applies, as here, then the expression given above is a completely general result; that is

$$\Delta H = E_{a,f} - E_{a,r} \qquad (25)$$

where $E_{a,f}$ is the activation energy for the forward reaction, and $E_{a,r}$ that for the corresponding reverse reaction.

5.4 THE RATE OF REACTION AND THE ACTIVATION ENERGY

We began Section 5 by saying that, in contrast to the equilibrium constant and the overall energy change, the rate of a reaction depends on *how* the reaction actually occurs at the molecular level, that is, on the route it takes—its mechanism. In Section 5.3 you met one path or mechanism for the simple gas reaction

$$Br(g) + H_2(g) \longrightarrow HBr(g) + H(g) \qquad (23)^*$$

Remember that the activation energy for this reaction is found experimentally to be about $82\,\mathrm{kJ\,mol^{-1}}$.

Compare this value with the bond dissociation energy for H_2, $D(\mathrm{H{-}H}) = 436\,\mathrm{kJ\,mol^{-1}}$. Can you draw any conclusions from this comparison?

On the one hand, these values show that the activation energy is much *less* than the energy required to break the H—H bond. This suggests that if the H—Br bond is forming *at the same time* as the H—H bond is breaking, the energy required is much less than that required to break the H—H bond alone. On the other hand, the experimental value of E_a immediately rules out a possible alternative mechanism, in which H_2 molecules are first completely dissociated into H atoms: such a mechanism requires E_a to be equal to $D(\mathrm{H{-}H})$, about five times as big as the experimental value. This is not surprising. The alternative mechanism suggested above would obviously be much more 'expensive' in energy terms, and it is generally true that a reaction will follow the path that involves the lowest energy barrier.

In summary: the mechanism, or path of the reaction, determines the height of the energy barrier; at a given temperature, the value of E_a has a strong influence on the speed of the reaction. It follows that a reaction will go faster if a mechanism involving a lower activation energy can be found. This can be achieved by performing the reaction in the presence of a **catalyst**, if one can be found. A catalyst is a substance that speeds up a reaction, usually without being consumed itself.

An example is the decomposition of hydrogen peroxide, H_2O_2, a reaction that you came across in SAQ 5. A concentrated aqueous solution of hydrogen peroxide is stable at ordinary temperatures. But add a few drops of an aqueous solution containing the iron(II) ion, $Fe^{2+}(aq)$, and the hydrogen peroxide rapidly decomposes to water and oxygen gas:

$$2H_2O_2(aq) = 2H_2O(l) + O_2(g) \qquad (26)$$

The presence of $Fe^{2+}(aq)$ speeds up the decomposition, even though it does not appear in the balanced equation for the reaction: Reaction 26 is *catalysed* by $Fe^{2+}(aq)$. A few drops of blood(!) have a similar effect, as does the addition of a lump of solid manganese dioxide (MnO_2). In the latter case, when Reaction 26 reaches equilibrium, the lump of MnO_2 remains, its mass unchanged. It speeds up the decomposition of H_2O_2 without being consumed itself: MnO_2 is also an effective catalyst for this reaction.

The study of catalysis is a subject in itself and, as the examples above suggest, the types of catalyst vary enormously. Later in this Course (Unit 22), you will meet some of the important biological catalysts, known collectively as *enzymes*. (It is an enzyme in your blood that catalyses the decomposition of hydrogen peroxide.) These are efficient (and very specific) catalysts, but they normally function only within a limited range of temperature.

Catalysts will not be discussed further in this Unit, other than to point out that the role of a catalyst is to increase the *rate* of a reaction, usually by changing the mechanism. In general, the resulting mechanism must have a lower activation energy than that without the catalyst if an increase in rate is to occur: this is illustrated schematically in Figure 14 for a hypothetical exothermic reaction, and symbolically in Figure 15. It follows from the picture in Figure 14 that a catalyst lowers the energy barrier for *both* the forward *and* the reverse reactions, and hence increases the rates of both. Thus, it does not affect the equilibrium position, but simply increases the rate at which this is attained: it *cannot* alter the value of the equilibrium constant for a reaction.

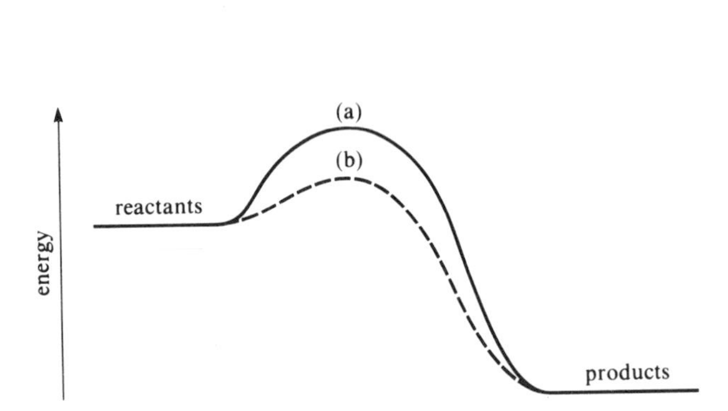

FIGURE 14 Schematic diagram to show the effect of a catalyst on the activation energy for a reaction: (a) without a catalyst; (b) in the presence of a catalyst.

FIGURE 15 A catalyst usually helps to lower the energy barrier between reactants and products.

There is one final question we should examine before closing this Section. Throughout, we have used as examples a few very simple gas reactions: how applicable are the ideas developed here to other, more complex, reactions? The most important point is that the balanced equation for a reaction does not automatically say anything about its mechanism: this can only be inferred from the *observed* influence of concentration, temperature, etc., on the actual rate of reaction. For example, the rate of the following, and apparently simple, gas reaction

$$H_2(g) + Br_2(g) = 2HBr(g)$$

is actually found to have a highly complex dependence on the concentrations of both reactants *and* product. The reason for this is that the reaction does not take place via simple collisions between H_2 and Br_2 molecules. Rather, it involves a series of simpler reactions, two of which have been used as examples in this Section (Equations 23 and 24); the steps are listed as follows:

$$Br_2 \longrightarrow Br + Br$$

$$Br + H_2 \longrightarrow HBr + H$$

$$H + Br_2 \longrightarrow HBr + Br$$

$$H + HBr \longrightarrow H_2 + Br$$

$$Br + Br \longrightarrow Br_2$$

Do not spend time trying to work out how this sequence of reactions eventually leads to the product, HBr: it is a little complicated. The sequence is included simply to illustrate the meaning of the terms 'mechanism' and 'steps' in a complex reaction. It is usually true, however, that for *each* of the steps in a complex reaction like this, there will be:

(a) a lowest energy path for reaction

(b) an energy barrier, and hence

(c) an activation energy.

SUMMARY OF SECTION 5

1 A simple collision model accounts in a qualitative way for the observed influence of concentration and temperature on the rate of a simple gas reaction.

2 A more quantitative treatment highlights certain limitations of this simple model. An examination of the discrepancies between the calculated and observed rate behaviour leads to a plausible modification of the simple collision model: a collision between reactant molecules leads to reaction only when its energy is at least as great as some threshold energy, known as the activation energy.

3 The activation energy can be pictured as an energy barrier that reactant molecules must surmount before products can be formed.

4 The height of the energy barrier, and hence the rate of a reaction, depends on the mechanism of the reaction—the path by which reactants are converted into products.

5 When a chemical reaction is inconveniently slow, it is sometimes possible to find a catalyst. This is a substance that speeds up the reaction, usually without being consumed itself. It changes the mechanism of the reaction, and lowers the energy barrier, without affecting the equilibrium position.

SAQ 8 State the deficiencies of a simple collision model for chemical reactions, and explain briefly how these deficiencies are overcome by the idea that reactant molecules must collide with sufficient energy (the activation energy) for reaction to occur.

SAQ 9 Consider the following gas reaction:

$$H(g) + Cl_2(g) \longrightarrow HCl(g) + Cl(g)$$

(a) What is the overall enthalpy change for the reaction? Is it endothermic or exothermic?

(b) Sketch a reaction-coordinate diagram for this reaction, and show ΔH and E_a on your sketch.

(c) What is the value of the activation energy E_a for this reaction?

SAQ 10 As you should have found in answering ITQ 6, the following reaction is highly exothermic:

$$2H_2(g) + O_2(g) = 2H_2O(g); \quad \Delta H = -482\,\text{kJ}\,\text{mol}^{-1}$$

Although the value of the equilibrium constant for this reaction is very large ($K = 3.3 \times 10^{81}\,\text{l mol}^{-1}$ at 25 °C), a gaseous mixture of hydrogen and oxygen shows no noticeable reaction under normal conditions. Can you suggest a possible reason?

However, if the mixture is passed over a mesh of platinum gauze, the reaction above proceeds rapidly. What is the role of the platinum gauze? Does its presence affect the overall enthalpy change, or the equilibrium position for the reaction?

6 REQUIREMENTS FOR REACTION TO OCCUR

Having studied Units 15 and 16, you have now met, in general terms, most of the factors that influence the progress of a chemical reaction. For a reaction to be observed to occur:

1 The reactant molecules must meet (collide);

2 The molecules must meet with sufficient energy to react (the activation energy);

3 The equilibrium constant must be sufficiently large to *allow* enough reaction to occur for it to be detected.

Requirement 1 is clearly affected by concentration; the more molecules there are in a given volume, the more likely they are to meet.

Requirement 2 depends on the mechanism(s) available to the reactants and, of course, on the temperature: this was the basis for the discussion in Section 5.4.

Requirement 3 cannot be altered at a given temperature, a fact that was emphasized in Section 9 of Unit 15.

But how does a *change* in the reaction conditions affect each of these requirements? First, increasing the concentration of the reactants increases the chance of requirement 1 being met, and hence generally increases the rate of the reaction. It also affects the equilibrium position. For example, consider again the reaction discussed in Section 7 of Unit 15:

$$H_2(g) + I_2(g) = 2HI(g) \tag{27}$$

for which

$$K = \frac{[HI(g)]^2}{[H_2(g)][I_2(g)]} \tag{28}$$

$$= 54.3 \text{ at } 427\,°C$$

Suppose that you have an equilibrium mixture of H_2, I_2 and HI at a constant temperature of 427 °C.

□ What will be the effect of adding more of one reactant, hydrogen say, to this mixture?

■ Addition of H_2 will *disturb* the equilibrium: $[H_2(g)]$ is increased, but the ratio of concentrations on the right-hand side of Equation 28 *cannot* change. As you saw in Unit 15, the system responds to this disturbance: more H_2 and I_2 react to form HI, thus maintaining the ratio of concentrations at its equilibrium value.

In terms of Le Chatelier's principle, the equilibrium in Equation 27 is shifted to the right, thereby increasing the equilibrium yield of HI.

□ Can you think of another way of displacing the equilibrium, and thus increasing the equilibrium yield of product(s) at constant temperature?

■ The simplest technique, in principle, is to remove product as it is formed. Again this does not constitute a change in K. (See the answers and comments to SAQs 8 and 19 in Unit 15.)

What about the effect of changing the temperature? Well, increasing the temperature increases the chance of reaction by meeting requirement 2 above: most reactions go faster at higher temperatures.

But how does a change in temperature affect the equilibrium position? It is time to return to the question which was left unanswered at the end of Unit 15 (Section 9). There we simply noted that the equilibrium constant for a reaction may either increase or decrease as the temperature is raised. You are now in a position to predict the way the equilibrium constant varies with temperature: it depends solely on the *sign* of the enthalpy change for the reaction, that is, on whether the reaction is endothermic or exothermic.

You can see that this connection is plausible by applying Le Chatelier's principle. In this case the external constraint is an increase in temperature. Now you saw in Section 2 that the initial temperature change resulting from a reaction depends on whether the reaction is exothermic or endothermic. An exothermic reaction leads to a temperature increase; an endothermic reaction to a temperature decrease. So when the external constraint is an increase in temperature, the constraint can be lessened if the equilibrium shifts in the direction in which the reaction is endothermic, because this shift would, by itself, tend to lower the temperature. Thus, if we write down a reaction that is endothermic in the left → right direction, the constraint imposed on the system by increasing the temperature will be reduced by the formation of more products, which corresponds to an *increase* in the value of the equilibrium constant. The reverse is true for a reaction that is exothermic in the left → right direction: an increase in temperature causes a *decrease* in K.

ITQ 7 The enthalpy change for the following reaction was given earlier:

$$H_2(g) + Cl_2(g) = 2HCl(g); \quad \Delta H = -184\,\text{kJ}\,\text{mol}^{-1}$$

How will the equilibrium constant for this reaction vary with temperature?

□ When will the effects of increasing temperature on requirements 2 and 3 above work in opposite directions?

■ When the reaction is exothermic. Then, increasing the temperature will reduce the size of K, and hence the maximum possible yield of product, no matter how much more quickly this is formed.

As you will see in the next Section, this dilemma is commonly encountered with slow exothermic reactions in the chemical industry: one can speed the reaction up by increasing the temperature, but only at the cost of reducing the yield. One way of escape is to search for a suitable catalyst, which will speed up the reaction (generally by decreasing the activation energy) without the temperature having to be raised unduly.

SUMMARY OF SECTION 6

1 For a reaction to be observed to occur:

(a) The reactant molecules must meet (collide);

(b) The molecules must meet with sufficient energy to react (the activation energy);

(c) The equilibrium constant must be sufficiently large to *allow* enough reaction to occur for it to be detected.

2 The rate of a reaction can generally be increased by increasing either the concentration(s) of reactant(s) or the temperature.

3 At a given temperature, the equilibrium yield of product can only be increased by displacing the equilibrium in some way.

4 The effect of temperature on the equilibrium constant (and hence equilibrium yield of product) depends on the sign of the enthalpy change for the reaction.

The following SAQ leads in to the discussion of the Haber process in the following Section, so make sure you try it.

SAQ 11 The Haber process for 'fixing' molecular nitrogen from the atmosphere centres around the manufacture of ammonia via the following reaction:

$$N_2(g) + 3H_2(g) = 2NH_3(g); \quad \Delta H = -93\,\text{kJ mol}^{-1}$$

where ΔH is the enthalpy change. At 25 °C the equilibrium constant for this reaction has the value $K = 2.7 \times 10^8\,\text{l}^2\,\text{mol}^{-2}$. But no appreciable reaction occurs at room temperature and pressure. In what ways could the yield of ammonia be increased?

7 THE HABER–BOSCH PROCESS

It seems fitting to close these two Units with a brief examination of an industrial process that not only has had a profound impact on our society, but also arose from a painstaking application of the scientific principles discussed in Units 15 and 16.

During the 19th century, the increase in world population was such that there was a dramatic increase in the demand for 'fixed' nitrogen, that is nitrogen in a chemically combined form, especially for use in fertilizers. By the turn of the century, it was recognized that the obvious place to start was the air, with its virtually unlimited supply of molecular nitrogen, free for the taking. However, the problem was more easily recognized than solved, for one of the main characteristics of molecular nitrogen (N_2) is its comparative unreactivity. As we said in Section 3.1, this can largely be attributed to the great strength of the N≡N bond in molecular nitrogen. Nitrogen is not as unreactive as the noble gases; indeed, many thousands of nitrogen compounds are known. Nevertheless, under ordinary conditions it reacts with very few substances. Thus, the problem was to induce the unreactive nitrogen to form a compound that could be a source of fixed nitrogen suitable for both agricultural and industrial purposes. One obvious candidate was ammonia, NH_3.

FIGURE 16 Fritz Haber (1868–1934) received a Nobel Prize for Chemistry in 1919 for work that led to the development of a process for the industrial synthesis of ammonia.

Despite its apparent simplicity, the direct synthesis of ammonia from the elements, Equation 16, proved elusive:

$$N_2(g) + 3H_2(g) \rightleftharpoons 2NH_3(g) \qquad (16)^*$$

Fritz Haber (Figure 16), who was Professor of Technical Chemistry at Karlsruhe in Germany, came to the ammonia problem in 1903. In spite of early setbacks and much contemporary criticism from the scientific community, Haber and his colleagues persevered. They carried out a systematic investigation of the reaction and its response to changing conditions. By 1909 they had developed the essential features of a practical process, and the main principles of their designs are still in use today—quite an achievement!

Having worked through SAQ 11, you will already have an idea of the basic requirements. To summarize: the value of the equilibrium constant at 25 °C ($K = 2.7 \times 10^8\,\text{l}^2\,\text{mol}^{-2}$) suggests that the equilibrium yield of ammonia should be appreciable: but the reaction is so slow at this temperature that it cannot be observed—it never reaches equilibrium. The rate of reaction can be increased by raising the temperature; but the reaction is exothermic ($\Delta H = -93\,\text{kJ mol}^{-1}$), so the maximum possible yield falls off with increas-

TABLE 5 Equilibrium yield of ammonia as a function of temperature

Temperature/°C	NH_3 at equilibrium (%)
100	79.6
200	29.4
300	6.0
400	1.5
500	0.5

ing temperature. In this particular case, the effect is quite dramatic, as you can see from Table 5: this gives the equilibrium yield of ammonia as a percentage from an initial mixture containing N_2 and H_2 in the ratio 1:3.

Thus, a compromise has to be reached between a reasonable rate of formation of ammonia, which requires a high temperature, and an acceptable equilibrium yield, which demands a low temperature. As mentioned earlier, this conflict is common in the chemical industry, but can often be resolved by the development of a suitable catalyst.

You have also seen that the reaction in Equation 16 can be driven in the desired direction by removing ammonia as it is formed. However, there is another, and crucially important, way of displacing an equilibrium like this.

In Units 13–14 you met *Avogadro's hypothesis*: 'equal volumes of all gases, under the same conditions of temperature and pressure, contain the same number of molecules.' A direct deduction from this hypothesis is that a mole (6×10^{23} molecules) of any gas, under the same conditions of temperature and pressure, must always have the same volume. It follows that at constant temperature and pressure, the volume on the left-hand side of Equation 16 must be twice that on the right-hand side:

$$\underset{1\ \text{vol}}{N_2(g)} + \underset{3\ \text{vols}}{3H_2(g)} \rightleftharpoons \underset{2\ \text{vols}}{2NH_3(g)} \qquad (16)^*$$

You are probably familiar with the idea a gas responds to the external constraint of an increase in the external pressure by contracting—by a decrease in volume.

□ Using Le Chatelier's principle, can you predict the effect of increasing pressure on the equilibrium in Equation 16?

■ According to Le Chatelier's principle, the influence of an *increase* in pressure will be lessened by a *reduction* in the volume. The equilibrium will shift to the side having the *smaller* volume—the right-hand side in this case.

In general, the side having the smaller number of moles of *gaseous* molecules is favoured by an increase in pressure. Notice that this does not constitute a change in the equilibrium constant: the latter depends *solely* on the temperature.

The effect of pressure on the equilibrium yield of ammonia is shown in Table 6: the corresponding values in Table 5 are for a pressure of one atmosphere (1 atm), effectively the pressure of the atmosphere at sea-level. Thus 25 atm, for example, represents a pressure about 25 times as great as that of the atmosphere. This was the extra factor that spelt success for Haber and his colleagues. Even in the presence of a catalyst, it was necessary to raise the temperature to a point at which the yield of ammonia was disappointingly poor at atmospheric pressure. The yield was improved to a commercially viable level by working at higher pressures. Even here, however, there is a need for compromise: although an increase in pressure favours the formation of ammonia, working on an industrial scale at high

TABLE 6 Equilibrium yield of ammonia as a function of temperature and pressure*

	NH_3 present at equilibrium (%)				
Temperature/°C	25 atm	50 atm	100 atm	200 atm	400 atm
100	91.7	94.5	96.7	98.4	99.4
200	63.6	73.5	82.0	89.0	94.6
300	27.4	39.6	53.1	66.7	79.7
400	8.7	15.4	25.4	38.8	55.4
500	2.9	5.6	10.5	18.3	31.9

* These values were taken from an article by S. P. S. Andrew, in *The Modern Inorganic Chemicals Industry* (1977), ed. R. Thompson, Special publication No. 31, The Chemical Society, London.

pressures involves engineering and technical problems, and the actual pressure employed is largely dictated by economic considerations.

Overall, then, the Haber process represents a series of informed compromises based on data obtained from many thousands of experiments. (In the commercial development of the process, led by the engineer, Carl Bosch, with the German company Badische Anilin und Soda Fabrik AG, some 6 500 experiments were carried out between 1910 and 1912 in order to find the most suitable catalyst!) Today, the process is run at a temperature in the range 400–540 °C and a pressure in the range 80–350 atm, in the presence of an activated iron catalyst containing small amounts of, typically, potassium, aluminium, silicon and magnesium oxides. In practice, the gases are circulated continually through a bed of the catalyst at such rates that the reaction does not reach equilibrium: the conversion per pass is generally quite low. However, the ammonia is condensed out of the gas stream, and the unchanged hydrogen and nitrogen are recirculated. The ammonia can then be combined with nitric or sulphuric acid to make the solid fertilizers, ammonium nitrate (NH_4NO_3) and ammonium sulphate ($(NH_4)_2SO_4$).

SUMMARY OF SECTION 7

1 The manufacture of nitrogen fertilizers depends upon the synthesis of ammonia from nitrogen and hydrogen, but at normal temperatures and pressures, the reaction is too slow.

2 Because the reaction is exothermic, too large an increase in temperature leads to an unacceptable lowering of the equilibrium yield.

3 With an iron catalyst, the rate is satisfactory at 400–540 °C, and so is the ammonia yield if the pressure of the gas mixture is raised to 80–350 atm.

SAQ 12 The nitrogen for the Haber process is freely available in the air, but what about the hydrogen? One source that is used in modern plants is natural gas (mainly methane), which reacts with steam as follows:

$$CH_4(g) + H_2O(g) \rightleftharpoons CO(g) + 3H_2(g)$$

(a) A Lewis structure for carbon monoxide is harder to draw than the examples of Units 13–14. However, it shows that there is a triple bond in the carbon monoxide molecule. Given that the bond energy of this C≡O bond is 1 076 kJ mol^{-1}, what is the enthalpy change for the above reaction?

(b) What will be the effect of changing temperature and pressure on the yield of hydrogen from this reaction?

SAQ 13 Consider the following gaseous equilibrium

$$2F_2(g) + O_2(g) \rightleftharpoons 2F_2O(g)$$

where $\Delta H = -43.5\,\text{kJ mol}^{-1}$.

(a) Write an expression for the equilibrium constant of this reaction. If concentrations are expressed in mol l^{-1}, what units does K have?

(b) What effect would the following changes have on the amount of F_2O present at equilibrium?

(i) O_2 is added.

(ii) The temperature is raised.

(iii) The volume of the container is reduced (but the total amount of material does not change).

(iv) An effective catalyst is added.

8 TV NOTES: ENERGY AND ROCKETS

This programme examines the criteria for a good rocket propellant—a particular combination of fuel plus oxidizer used for rocket propulsion. The main emphasis is on the question 'What makes a reaction exothermic?', and the treatment aims to reinforce that in Sections 3.1 and 3.2.

The programme began with Dr Logan demonstrating the spectacular, and obviously exothermic, reactions of dinitrogen tetroxide (N_2O_4) with two other substances—aniline and unsymmetrical dimethylhydrazine (UDMH for short). As Dr Logan's model suggested, the second reaction (N_2O_4 + UDMH) has indeed been used for propulsion, chiefly in the small lateral rocket motors used to adjust the orbit of the lunar module during the Apollo Moon shots.

We then went on to examine the three general properties that any good rocket propellant must have. In summary, these are as follows:

1 *The reaction between the components must be highly exothermic* (That is, ΔH should be large and negative). We used the following simple example:

$$H_2(g) + F_2(g) = 2HF(g) \qquad (29)$$

to establish that a reaction will be exothermic if it results in the formation of strong bonds at the expense of breaking weaker ones.

As mentioned above, the meaning and use of bond energies are discussed in more detail in Section 3, and a parallel discussion of more conventional chemical fuels is given in Section 4. Don't worry if the actual calculation of the overall energy change for Reaction 29 seemed a bit too quick: it is repeated in ITQ 4.

2 *The reaction between the components must be fast.* We deduced this criterion from the great simplicity of the 'business end' of a rocket motor—the combustion chamber and exhaust nozzle (see Figures 17 and 18, for example.) The combustion chamber is just a hollow tube, and the reactants would simply be swept away if the reaction between them were not effectively instantaneous. All of the reactions you saw certainly fulfilled this criterion, as well as the one above.

FIGURE 17 The Walter 509 rocket engine from the Messerschmidt 163: (a) pumping mechanism for delivering fuel and oxidizer to the combustion chamber; (b) combustion chamber; (c) exhaust nozzle.

FIGURE 18 End-on view of the Walter 509 engine, showing inlet valves to the combustion chamber.

3 *The products of the reaction should be gases of low molar mass.** In a rocket engine, chemical energy is converted *directly* into kinetic energy of the product molecules in the combustion reaction. The net result is a stream

* Please forgive the anachronism of 'molecular weight' for 'molar mass' that slipped in towards the end of the programme!

of hot, fast-moving gases that are ejected through the exhaust nozzle in one direction. The 'reaction' to this causes the body of the rocket to move in the opposite direction, for much the same reason as a rifle 'kicks back' when fired (Figure 19).

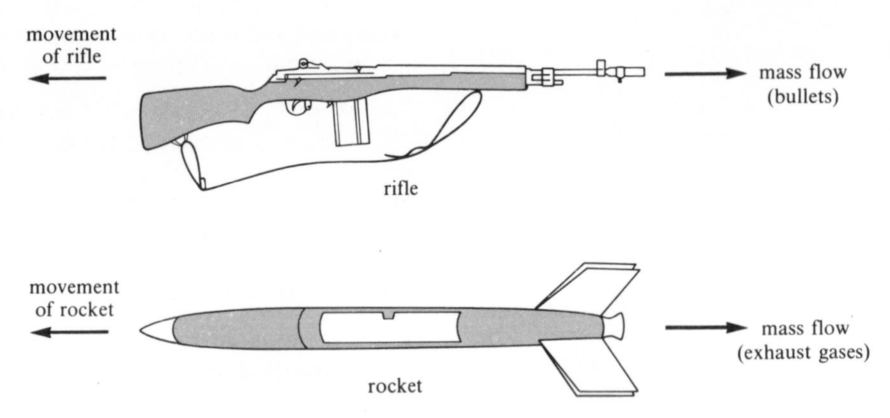

FIGURE 19 A comparison of the motion of a rocket and the recoil of a rifle. In both instances momentum is conserved: the momentum of the bullet (exhaust gases) is equal in magnitude, but of opposite sign, to the momentum of the rifle (rocket).

Although we did not discuss it in the programme, this is a further example of the law of conservation of momentum, which you met in Unit 3. Thus, if the rocket is stationary when the engine is fired,

$$\text{momentum of rocket} = -\text{momentum of exhaust gases}$$
$$= -m_e v_e$$

where m_e is the mass and v_e the speed of the exhaust gases. As the minus sign implies, the rocket moves in the opposite direction to the exhaust stream.

But a rocket has to carry its own propellant, so we obviously want as large a momentum as possible per unit mass of propellant burned. In other words, one important criterion for a good rocket propellant is the size of the exhaust speed, v_e.

The simple demonstration you saw in the programme showed that, under identical conditions, a balloon of hydrogen (molar mass = $2\,\text{g mol}^{-1}$) emptied a lot faster than one containing the same amount of carbon dioxide (molar mass = $44\,\text{g mol}^{-1}$). In other words, the lower the molar mass, the higher the exhaust speed.

The final part of the programme showed these criteria in action in one of the forerunners of today's interplanetary rockets—the Messerschmidt 163. The propellant was as follows:

oxidizer (T-Stoff): an 80% aqueous solution of hydrogen peroxide (H_2O_2)

fuel (C-Stoff): a solution of hydrazine (N_2H_4) in methanol (methyl alcohol)

The reaction between these components produces nitrogen and steam:

$$N_2H_4 + 2H_2O_2 = N_2 + 4H_2O \tag{30}$$

or, using dashes to represent bonds,

$$\underset{\displaystyle\;\;\;\text{H}\;\;\;\text{H}}{\text{H—}\underset{|}{\text{N}}\text{—}\underset{|}{\text{N}}\text{—H}} + 2\,\text{H—O—O—H} \longrightarrow \text{N}\equiv\text{N} + 4\,\text{H—O—H}$$

We showed that this particular combination does have the three properties outlined above:

1 In Reaction 30, some weak bonds (especially O—O and N—N) disappear, and they are replaced by strong ones (N≡N and O—H): the reaction is highly exothermic. The calculation of ΔH for this reaction is part of SAQ 7.

2 You saw that the reaction is very fast.

3 The products have relatively low molar masses: both are less dense than air.

OBJECTIVES FOR UNIT 16

After you have worked through this Unit, you should be able to:

1 Explain the meaning of, and use correctly, all the terms flagged in the text.

2 Given the value of the enthalpy change for a reaction, decide whether the reaction is endothermic or exothermic, and hence whether heat will be absorbed or released at constant temperature. (*ITQ 4; SAQs 1, 5, 7, 9, 11, 12 and 13*)

3 Decide whether a given phase change is either exothermic or endothermic. (*ITQ 2; SAQs 2 and 3*)

4 Draw Lewis structures (or recognize correct structures) for simple gas molecules and hence list the bonds broken and formed in a given gas reaction. (*ITQs 3, 5 and 6; SAQ 4*)

5(a) Given a Table of bond energies or average bond energies, together with the information in Objective 4, calculate the enthalpy change for a simple gas reaction. (*ITQs 4 and 6; SAQs 4, 5, 9 and 12*)

(b) Using Hess's law and appropriate enthalpies for phase changes, extend the procedure in (a) to the calculation of enthalpy changes for reactions involving liquids or solids. (*SAQs 4 and 5*)

6 Explain the limitations of the procedure outlined in Objective 5. (*SAQ 6*)

7 Use the procedure outlined in Objectives 4, 5 and 6 to discuss:

(a) the combustion of a typical chemical fuel in oxygen;

(b) the reaction of a typical fuel and oxidizer used for rocket propulsion. (*TV; SAQs 4 and 7*)

8 Explain in general terms how a collision model of chemical reactions, together with the idea of a threshold or activation energy, accounts for the observed influence of concentration and temperature on the rate of a reaction. (*SAQ 8*)

9(a) Sketch (or recognize a correct sketch of) the reaction-coordinate diagram for a simple reaction of given ΔH.

(b) Using such a sketch, distinguish between the enthalpy change for a reaction and its activation energy. (*SAQ 9*)

10 Relate the role of a catalyst to a reaction-coordinate diagram. (*SAQs 10 and 13*)

11 Describe, in terms of Le Chatelier's principle, how the equilibrium constant varies with temperature for (a) exothermic reactions, and (b) endothermic reactions. (*ITQ 7; SAQs 11, 12 and 13*)

12 List the requirements for a chemical reaction to occur and hence:

(a) predict the effect of changing reaction conditions (for example, temperature and pressure) on the progress of a given reaction;

(b) show how these factors influence the choice of reaction conditions for a given industrial process. (*SAQs 11, 12 and 13*)

ITQ ANSWERS AND COMMENTS

ITQ 1 1 200 kJ. You want to vaporize 500 cm^3, that is 500 g, of water.

The molar mass of H_2O is $(2 + 16)\,g = 18\,g$.

So 500 g represents $\frac{500}{18}$ mol of H_2O.

To vaporize 1 mol requires 43.3 kJ.

So $\frac{500}{18}$ mol require $\frac{500}{18} \times 43.3\,kJ \approx 1\,200\,kJ$.

ITQ 2 Yes. The endothermic process ether(l) → ether(g) takes energy from its surroundings, including your skin, and so leaves you feeling colder.

ITQ 3 As you saw in Units 13–14, nitrogen has five electrons in its outermost shell. In N_2 each nitrogen atom can attain a noble gas structure by sharing *three* pairs of electrons:

$$\overset{\times}{\underset{\times}{}}N \overset{\times\times\times}{\underset{\times\times\times}{}} N\overset{\times}{\underset{\times}{}} \qquad \text{or, more simply,} \qquad N{\equiv}N$$

Similarly, in NH_3, all the atoms can attain noble gas structures by sharing pairs of electrons:

$$\begin{array}{c} \times\times \\ H \overset{\times}{\bullet} N \overset{\bullet}{\times} H \\ \times\bullet \\ H \end{array} \qquad \text{or} \qquad \begin{array}{c} H{-}N{-}H \\ | \\ H \end{array}$$

(Remember, structures like this are simply a representation of the bonds present in the molecule, and do not reflect the actual shape of the molecule.) Thus, in the reaction given,

$$3H_2(g) + N_2(g) = 2NH_3(g)$$

three H—H bonds and one N≡N bond are broken for every six N—H bonds formed.

ITQ 4 In the reaction

$$H_2(g) + F_2(g) = 2HF(g)$$

one H—H and one F—F bond are broken for every two H—F bonds formed. The reaction is very similar to the one discussed in the text. Thus

$$\Delta H = D(H{-}H) + D(F{-}F) - 2D(H{-}F)$$

$$= (436 + 158 - 2 \times 568)\,kJ\,mol^{-1}$$

$$= -542\,kJ\,mol^{-1}$$

The reaction is strongly exothermic, largely because it results in the formation of very strong H—F bonds at the expense of breaking weaker ones. Indeed, the weakness of the F—F bond suggests why many reactions with fluorine are strongly exothermic (see Section 4).

ITQ 5 Oxygen has six electrons in its outermost shell, and hydrogen has one, so in H_2O_2 each atom can attain a noble gas electronic configuration by sharing pairs of electrons, as:

$$\begin{array}{c} \times\times\ \ \times\times \\ H \overset{\times}{\bullet} O \overset{\times}{\times} O \overset{\times}{\bullet} H \\ \times\times\ \ \times\times \end{array} \qquad \text{or} \qquad H{-}O{-}O{-}H$$

Similarly, the N and H atoms in N_2H_4 can attain noble gas electronic configurations as follows:

$$\begin{array}{c} H\ \ H \\ \times\bullet\ \ \times\bullet \\ \overset{\times}{\times} N \overset{\times}{\times} N \overset{\times}{\times} \\ \bullet\times\ \ \times\bullet \\ H\ \ H \end{array} \qquad \text{or} \qquad \begin{array}{c} H\ \ H \\ |\ \ \ | \\ N{-}N \\ |\ \ \ | \\ H\ \ H \end{array}$$

ITQ 6 As in ITQ 5, in H_2O each atom can attain a noble gas electronic configuration by sharing pairs of electrons, as:

$$\begin{array}{c} \times\times \\ H \overset{\times}{\bullet} O \overset{\times}{\bullet} H \\ \times\times \end{array} \qquad \text{or} \qquad H{-}O{-}H$$

In the reaction given,

$$2H_2(g) + O_2(g) = 2H_2O(g)$$

two H—H bonds and one O=O bond (in O_2) are broken for every four O—H bonds formed. Thus

$$\Delta H = 2D(H{-}H) + D(O{=}O) - 4D(O{-}H)$$

Tables 1 and 3 include values for these bond energies, whence

$$\Delta H = (2 \times 436 + 498 - 4 \times 463)\,kJ\,mol^{-1}$$

$$= -482\,kJ\,mol^{-1}$$

The heat released is 482 kJ per mole of the reaction *as written*, that is, per two moles of hydrogen and one mole of oxygen consumed, and two moles of water vapour formed.

ITQ 7 ΔH is negative: the reaction is exothermic, so the equilibrium constant should *decrease* with increasing temperature. Le Chatelier's principle leads to this conclusion if you think of the increase in temperature as an external constraint. This constraint can be reduced if the equilibrium shifts in the endothermic direction—towards the left-hand side (reactants). This corresponds to a *decrease* in K.

SAQ ANSWERS AND COMMENTS

SAQ 1 (a) (ii) and (iii) are true.

The enthalpy change in a reaction is given by the sum of the enthalpies of the products less the enthalpies of the reactants (Equation 7, Section 2). If the sum of the enthalpies of the products is less than that of the reactants, ΔH will be negative. By the law of conservation of energy, the energy lost must go somewhere, and it results in heat being evolved, at constant temperature.

(b) From the reaction *as written*, complete reaction of 1 mol (2 g) of H_2 with 1 mol (2 × 80 g) of Br_2 at constant temperature releases 72 kJ of heat. Thus, complete reaction between 0.5 mol of each must release 36 kJ.

SAQ 2 (a) This process will be endothermic, since the transition from solid to gas absorbs energy.

(b) This process represents the dissociation of solid sodium chloride into sodium ions and chloride ions in aqueous solution. It is not possible to decide whether such a reaction is exothermic or endothermic on the basis of the simple generalizations given in the text.

(c) You should recognize from everyday experience that this process, the combustion of methane (natural gas) in oxygen, is highly exothermic.

(d) This process is freezing and is the *reverse* of the endothermic change from $H_2O(s)$ to $H_2O(l)$ (that is melting). By the law of energy conservation, this must be exothermic, releasing an equivalent amount of energy.

SAQ 3 Boiling water (100 °C) is at a much higher temperature than your body (37 °C). In contact with your skin, the water cools down, transferring heat to your skin and damaging the tissues. A scald from steam is usually more serious, because the steam first condenses to water. According to the discussion in Section 2, this is an *exothermic* process, releasing heat. It is this extra heat, over and above that transferred as the water subsequently cools down, that does the extra damage. Also, of course, steam usually damages a much larger area of skin than the same mass of boiling water, because steam is much less dense.

SAQ 4 (a) As you saw in Units 13–14, carbon has four electrons in its outermost shell, whereas oxygen has six. In the CO_2 molecule, each oxygen atom can attain a noble gas electronic configuration by sharing two pairs of electrons with carbon:

$$:\!\ddot{O}\!::\!C\!::\!\ddot{O}\!: \quad \text{or} \quad O{=}C{=}O$$

At the same time, of course, the carbon atom also attains a complete shell.

The reaction of interest is

$$C(g) + O_2(g) = CO_2(g) \qquad (31)$$

in which one O=O bond is broken for every two C=O bonds formed.

(b) Notice that in order to calculate ΔH for this reaction you do not need to know the 'energy' of C(g): only the changes in *bonding* are important. Thus

$$\Delta H = D(O{=}O) - 2D(C{=}O)$$

Taking values from Tables 1 and 3 gives

$$\Delta H = (498 - 2 \times 804)\,\text{kJ mol}^{-1}$$
$$= -1110\,\text{kJ mol}^{-1}$$

The value of ΔH given in Section 2 refers to the reaction

$$C(s) + O_2(g) = CO_2(g); \quad \Delta H = -393.5\,\text{kJ mol}^{-1} \qquad (1)^*$$

The discrepancy between the two values relates to the different physical states of carbon involved in the two reactions. Indeed, Hess's law can be used to add the *reverse* of Equation 31 to Equation 1, giving the following:

$$C(s) + O_2(g) = CO_2(g) \qquad \Delta H = -393.5\,\text{kJ mol}^{-1}$$
$$CO_2(g) = C(g) + O_2(g) \qquad \Delta H = +1110\,\text{kJ mol}^{-1}$$
$$C(s) = C(g) \qquad \Delta H = +716.5\,\text{kJ mol}^{-1}$$

This shows that the change from solid carbon to gaseous carbon is an endothermic process, as would be expected on the basis of the generalization in Section 2.

SAQ 5 From ITQs 5 and 6, H_2O and H_2O_2 can be represented as H—O—H and H—O—O—H, respectively. So the reaction of interest,

$$2H_2O_2(g) = 2H_2O(g) + O_2(g)$$

involves breaking two O—O and four O—H bonds for every four O—H and one O=O bonds made. Since the number of O—H bonds broken is the same as the number formed, these need not be considered. Thus

$$\Delta H = 2D(O{-}O) - D(O{=}O)$$

Taking values from Tables 1 and 3,

$$\Delta H = (2 \times 143 - 498)\,\text{kJ mol}^{-1}$$
$$= -212\,\text{kJ mol}^{-1}$$

The fact that this reaction is exothermic can be wholly attributed to the weakness of the O—O single bond in H_2O_2 compared with the double bond in O_2.

(b) $\Delta H = -195.4\,\text{kJ mol}^{-1}$. You have the following information:

$$2H_2O_2(g) = 2H_2O(g) + O_2(g) \quad \Delta H = -212\,\text{kJ mol}^{-1} \qquad (32)$$

$$H_2O(l) = H_2O(g) \qquad \Delta H = +43.3\,\text{kJ mol}^{-1} \qquad (33)$$

$$H_2O_2(l) = H_2O_2(g) \qquad \Delta H = +51.6\,\text{kJ mol}^{-1} \qquad (34)$$

Then the reaction of interest

$$2H_2O_2(l) = 2H_2O(l) + O_2(g); \quad \Delta H = ?$$

is the *sum* of Equation 32 and twice Equation 34 minus twice Equation 33. According to Hess's law, the values of ΔH must be related in the same way.

Reaction	ΔH/kJ mol^{-1}
$2H_2O_2(g) = 2H_2O(g) + O_2(g)$	-212
$2H_2O_2(l) = 2H_2O_2(g)$	$+2 \times 51.6 = +103.2$
$2H_2O(g) = 2H_2O(l)$	$-2 \times 43.3 = -86.6$
$2H_2O_2(l) = 2H_2O(l) + O_2(g)$	$\Delta H = -195.4$ kJ mol^{-1}

SAQ 6 The reaction

$$HCl(g) = H^+(aq) + Cl^-(aq)$$

represents the solution and dissociation of HCl in water. This process does *not* correspond to the breaking of a simple covalent bond: the bond dissociation energy of HCl represents the energy required for the following process:

$$HCl(g) = H(g) + Cl(g)$$

SAQ 7 From ITQ 5, H_2O_2 and N_2H_4 can be represented as:

H—O—O—H and $\begin{matrix} H & & H \\ | & & | \\ N & — & N \\ | & & | \\ H & & H \end{matrix}$

The bonds broken and formed in the reaction

$$2H_2O_2(g) + N_2H_4(g) = 4H_2O(g) + N_2(g)$$

are listed in Table 7, together with values of D from Tables 1 and 3.

So $\Delta H = (3\,861 - 4\,649)$ kJ mol^{-1}

$= -788$ kJ mol^{-1}

You can see from Table 7 that O—H and N—H bonds are of roughly comparable strength, and the same total numbers are broken and made. Thus, the exothermic nature of this reaction can be largely attributed to the weakness of the N—N and O—O bonds (broken) and the strength of the N≡N bond (made).

SAQ 8 You are referred to Sections 5.1–5.3; the most important points are as follows:

1 A simple collision model predicts a dependence of reaction rate on concentration and on temperature, which agrees in kind with the observed behaviour. However, this simple model overestimates the rate of reaction at a given temperature.

2 The idea of an energy barrier to reaction overcomes this deficiency because it provides a plausible explanation of why a proportion of collisions do not lead to reaction.

SAQ 9 The reaction of interest is

$$H(g) + Cl_2(g) \longrightarrow HCl(g) + Cl(g)$$

(a) In this reaction, one Cl—Cl bond is broken for every H—Cl bond formed so

$$\Delta H = D(\text{Cl—Cl}) - D(\text{H—Cl})$$
$$= (244 - 432)\,\text{kJ mol}^{-1}$$
$$= -188\,\text{kJ mol}^{-1}$$

As ΔH is negative, the reaction is exothermic.

(b) The energy profile will have the general shape shown in Figure 20.

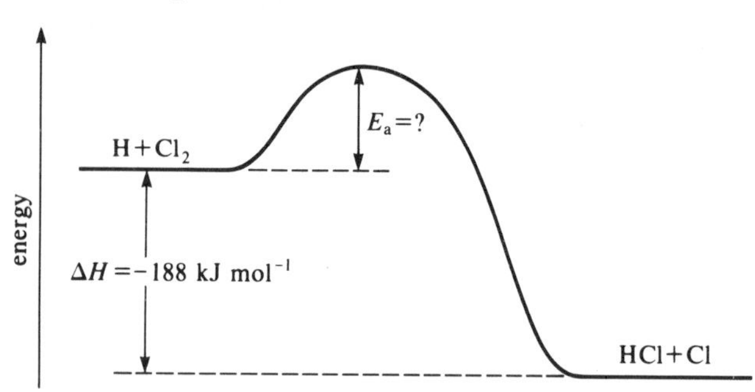

FIGURE 20 Reaction-coordinate diagram for the reaction $H(g) + Cl_2(g) \longrightarrow HCl(g) + Cl(g)$.

(c) This question cannot be answered from the information given. From the discussion in Section 5.4, you could perhaps place an upper limit on E_a: this would be D(Cl—Cl), that is 244 kJ mol^{-1}. However, the corresponding values for

$$Br(g) + H_2(g) \longrightarrow HBr(g) + H(g); \quad E_a = 82\,\text{kJ mol}^{-1} \quad (23)^*$$

and

$$H(g) + HBr(g) \longrightarrow H_2(g) + Br(g); \quad E_a = 12\,\text{kJ mol}^{-1} \quad (24)^*$$

suggest that the activation energy may be well below this upper limit. It's actually about 10 kJ mol^{-1}.

TABLE 7

Bonds broken			Bonds formed		
No.	Bond	D/kJ mol^{-1}	No.	Bond	D/kJ mol^{-1}
4	O—H	4 × 463 = 1 852	8	O—H	8 × 463 = 3 704
4	N—H	4 × 391 = 1 564	1	N≡N	1 × 945 = 945
1	N—N	1 × 159 = 159			
2	O—O	2 × 143 = 286			4 649
		3 861			

SAQ 10 The size of the equilibrium constant suggests that the *equilibrium* position strongly favours formation of product at 25 °C. The lack of reaction must be because the rate is very low, and this suggests that the activation energy is probably high.

The platinum gauze acts as a catalyst: it changes the mechanism of the reaction, lowering the energy barrier and hence increasing the speed of the reaction. According to the simple picture in Figure 14 (Section 5.4), a catalyst affects neither the overall energy change ΔH nor the final equilibrium position; that is, the value of K does not change.

SAQ 11 As in SAQ 10, the value of K at 25 °C is large, so the equilibrium position favours the formation of ammonia at this temperature. Again, the rate of reaction must be slow under normal conditions: the activation energy is high.

The rate of reaction can be increased by raising the temperature: but there is a problem. The reaction is exothermic, so K will decrease with increasing temperature: the temperature chosen will be a compromise between the need for a reasonable rate and an acceptable equilibrium yield.

If an efficient catalyst can be found, this will speed up the reaction (by lowering the energy barrier) without the temperature having to be raised unduly.

Finally, the reaction could be driven in the forward direction by removing the ammonia as it is formed. You deduced this result in answering SAQs 8 and 19 of Unit 15.

SAQ 12 (a) The reaction of interest is

$$CH_4(g) + H_2O(g) \rightleftharpoons CO(g) + 3H_2(g)$$

in which four C—H and two O—H bonds are broken for every one C≡O bond and three H—H bonds formed. Thus

$$\Delta H = 4D(\text{C—H}) + 2D(\text{O—H}) - D(\text{C≡O}) - 3D(\text{H—H})$$

Taking values from Tables 1 and 3, and using the one given,

$$\Delta H = (4 \times 416 + 2 \times 463 - 1076 - 3 \times 436)\,\text{kJ mol}^{-1}$$

$$= +206\,\text{kJ mol}^{-1}$$

(b) The reaction is endothermic, so both the rate of reaction and the equilibrium yield of product will be increased by raising the temperature.

From the reaction equation and Avogadro's hypothesis, the volume on the right-hand side is twice that on the left. According to Le Chatelier's principle, the effect of an increase in pressure will drive the equilibrium over to the side with the smaller volume, the reactant side. So to increase the yield of hydrogen the reaction should ideally be run under low pressure.

SAQ 13 (a) $K = \dfrac{[F_2O(g)]^2}{[F_2(g)]^2[O_2(g)]}$

and the units of K are

$$\frac{(\text{mol l}^{-1})^2}{(\text{mol l}^{-1})^2(\text{mol l}^{-1})} = \frac{1}{\text{mol l}^{-1}} = \text{l mol}^{-1}$$

If you had difficulty with this question, refer again to Section 7 of Unit 15.

(b) (i) If O_2 is added, the equilibrium is disturbed. According to Le Chatelier's principle, balance can be restored if the equilibrium shifts to the right, thus increasing $[F_2O(g)]$.

(ii) The reaction is exothermic, so increasing the temperature will reduce the value of K and hence the equilibrium yield of F_2O.

(iii) If the *total* amount of material is constant, reducing the volume of the container increases the overall pressure. According to Le Chatelier's principle, the effect of an increase in pressure is to shift the equilibrium to the side with the smaller number of gaseous molecules, the right-hand side in this case. This increases the equilibrium yield of F_2O.

(iv) Provided the system is at equilibrium, adding a catalyst can have no effect on the yield of F_2O: a catalyst alters the rate of a reaction but not the equilibrium position.

If you had difficulty with these questions, refer again to Section 4.4 of Unit 15, and Sections 6 and 7 of Unit 16.

ACKNOWLEDGEMENTS

Grateful acknowledgement is made to the following sources for permission to reproduce illustrations in this Unit:

Figure 1 Keystone Press: *Figure 8* Billings Energy Corporation; *Figure 16* courtesy of Dr L. Haber.

We would like to thank Dr I. E. Smith of the Cranfield Institute for his help with the TV programme, and Dr J. Holloway of the University of Leicester for setting up the hydrogen/fluorine demonstration.

INDEX FOR UNIT 16